Occupational Medicine

Firefighters' Safety and Health

Guest Editors:

Peter Orris, MD, MPH
Division of Occupational Medicine
Cook County Hospital
Chicago, Illinois

James Melius, MD, DrPH
Center to Protect Workers' Rights
Washington, DC

Richard M. Duffy, MS
International Association of Fire Fighters
Washington, DC

STATE OF THE ART REVIEWS

Publisher: HANLEY & BELFUS, INC.
             210 South 13th Street
             Philadelphia, PA 19107
             (215) 546-4995
             Fax (215) 790-9330

OCCUPATIONAL MEDICINE: State of the Art Reviews is included in *Index Medicus, MEDLINE, BioSciences Information Service, Current Contents and ISI/BIOMED.*

Authorization to photocopy items for internal or personal use, or the internal or personal use of specific clients, is granted by Hanley & Belfus, Inc. for libraries and other users registered with the Copyright Clearance Center (CCC) Transaction Reporting Service, provided that the base fee of $1.00 per copy plus $0.25 per page is paid directly to the CCC, 21 Congress St., Salem, MA 01970. Identify this publication by including with your payment the fee code, 0885-114X/95 $1.00 + .25.

OCCUPATIONAL MEDICINE: State of the Art Reviews   (ISSN 0885-114X)
October–December 1995   Volume 10, Number 4   (ISBN 1-56053-182-7)

©1995 by Hanley & Belfus, Inc. under the International Copyright Union. All rights reserved. No part of this book may be reproduced, reused, republished, or transmitted in any form or by any means without written permission of the publisher. An exception is chapters written by employees of U.S. government agencies; these chapters are public domain.

OCCUPATIONAL MEDICINE: State of the Art Reviews is published quarterly by Hanley & Belfus, Inc., 210 South 13th Street, Philadelphia, Pennsylvania 19107. Second-class postage paid at Philadelphia, PA, and at additional mailing offices.

POSTMASTER: Send address changes to OCCUPATIONAL MEDICINE: State of the Art Reviews, Hanley & Belfus, Inc., 210 South 13th Street, Philadelphia, PA 19107.

The 1995 subscription price is $86.00 per year U.S., $96.00 outside U.S. (add $40.00 for air mail).

Occupational Medicine: State of the Art Reviews
Vol. 10, No. 4, October–December 1995

FIREFIGHTERS' SAFETY AND HEALTH
Peter Orris, MD, MPH
James Melius, MD, DrPH, and
Richard M. Duffy, MS, Editors

CONTENTS

Preface xi
Peter Orris, James Melius, and Richard M. Duffy

Combustion Products and Other Firefighter Exposures 691
Peter S.J. Lees

> The author examines a wide variety of substances, from smoke, to synthetic products, and even the products that are used to fight fires. Measurements of exposure are discussed for carbon dioxide, carbon monoxide, hydrogen chloride, hydrogen cyanide, formaldehyde, acetaldehyde, benzene, and other chemicals.

Burn Injuries in Firefighters 707
Victor G. Cimino, Seth M. Krosner, and Marella L. Hanumadass

> The authors cover the care of burn injuries from start to finish, beginning with a discussion of immediate intervention and concluding with a look at psychosocial aspects of burns. Topics in the middle include early management, evaluation of the patient and classification of the burn's severity, burn resuscitation, the pathophysiology of smoke inhalation, dressing of burn wounds, escharotomies and fasciotomies, surgical management, and rehabilitation.

Smoke Inhalation Among Firefighters 721
Kenneth E. Bizovi and Jerrold D. Leikin

> Smoke inhalation may account for up to 75% of fire-related deaths and presents with a wide variety of complaints and findings. The authors examine the components of smoke to illustrate the patterns of smoke injury, provide useful guidelines on evaluation and management, survey current laboratory and diagnostic studies, and present their recommendations for treatment.

**Musculoskeletal Injury: Ergonomics
and Physical Fitness in Firefighters** 735
Paul A. Reichelt and Karen M. Conrad

> As a means of reducing the increasing incidence of musculoskeletal injuries in firefighters, the authors offer a framework for a program that would integrate hazard control and health promotion approaches. Particular focus is placed on the role of ergonomics and physical fitness factors in preventing these injuries.

Communicable Disease and Firefighters 747
Virginia M. Weaver and Sharon Doyle Arndt

> As firefighters are increasingly working in emergency medical response roles, their exposure to infectious diseases has also increased. This article discusses the various pathogens to which firefighters are particularly exposed and outlines measures to reduce their incidence. Legislation and resources specific to the fire service also are identified.

Occupational Stress in Contemporary Fire Service 763
Richard Gist and S. Joseph Woodall

> Although it is universally accepted that firefighting ranks among the most stressful occupations in the U.S., the quality of information on the matter has often been limited and subject to extreme variation. To help address the situation, the authors provide a thorough review of the literature on psychic trauma resulting from critical occupational events, draw specific applications to fire and rescue enterprises, discuss models of addressing stress disorders, and argue for a reasoned, ethical, and effective approach to the problem.

Pulmonary Effects of Firefighting 789
Cornelius H. Scannell and John R. Balmes

> The authors examine the acute and chronic effects of exposure to smoke among firefighters and look at mortality studies for the risk of death due to nonmalignant respiratory disease and lung cancer.

The Risk of Cancer in Firefighters 803
Anne L. Golden, Steven B. Markowitz, and Philip J. Landrigan

> A substantial body of literature now exists on the carcinogenic hazards of firefighting. The authors discuss in detail the data on the carcinogens benzene, asbestos, PAHs, formaldehyde, and diesel exhaust, and they go on to examine the prevalent cancers in firefighters, including leukemia, nonHodgkin's lymphoma, multiple myeloma, and cancer of the brain and bladder.

Cardiovascular Disease Among Firefighters 821
James M. Melius

> The author reviews the literature of the past 20 years on heart disease among firefighters, covering the specific aspects of firefighting that may be related to potential cardiovascular disease. The author then outlines steps that can be taken to reduce the risks of developing cardiovascular disease.

Reproductive Hazards and Firefighters 829
Melissa McDiarmid and Jacqueline Agnew

> The authors summarize the available data on three populations at potential risk from reproductive toxins: men, women, and developing fetuses. Among the areas discussed are the mechanisms of reproductive toxicity, industrial hygiene in the firefighting environment, and chemical and nonchemical reproductive hazards.

Noise and Hearing Loss in Firefighting **843**
Randy L. Tubbs

> Since the NIH received a request to investigate the high degree of hearing loss in a fire department in 1980, hearing loss among firefighters has become an area of increased investigation. The author identifies the sources of occupational noise in firefighting, looks at audiometric testing and recent research in firefighting noise, and presents guidelines for implementing hearing conservation programs.

Respiratory Health Hazards and Lung Function in Wildland Firefighters **857**
Robert Harrison, Barbara L. Materna, and Nathaniel Rothman

> The authors discuss the multitude of contaminants to which wildland firefighters are exposed, including carbon monoxide, sulfur dioxide, particulate and silica, polyaromatic hydrocarbons, aldehydes, and benzene. They examine the respiratory effects of these contaminants and then present their recommendations for an occupational health program for wildland firefighters.

Firefighters: Fitness for Duty **871**
Dick Gerkin

> One of the functions of the fire department physician is to give candidates and firefighters clearance to work based on medical and physical performance criteria. In this chapter the author discusses the practical application of the medical standards of NFPA 1582 and physical standards as measured by agility tests and the proposed NFPA 1583, presenting guidelines for hiring, return to work, and administration of periodic exams.

Index **877**

CONTRIBUTORS

Jacqueline Agnew, RN, PhD
Department of Environmental Health Sciences, The Johns Hopkins University School of Hygiene and Public Health, Baltimore, Maryland

Sharon Doyle Arndt, MPH
Safety and Health Assistant, Department of Occupational Health and Safety, International Association of Fire Fighters, Washington, DC

John R. Balmes, MD
Associate Professor of Medicine, Division of Occupational and Environmental Medicine and Pulmonary Medicine, University of California at San Francisco, San Francisco, California

Kenneth E. Bizovi, MD, ABEM
Division of Occupational Medicine, Cook County Hospital, Chicago, Illinois

Victor G. Cimino, MD, DDS
Attending Surgeon, Division of Burn Services, Department of Trauma, Cook County Hospital, Chicago, Illinois

Karen M. Conrad, PhD, RN, MPH
Associate Professor, Department of Public Health, Mental Health, and Administrative Nursing, College of Nursing, University of Illinois at Chicago, Chicago, Illinois

Richard M. Duffy, MS
Director of Health and Safety, International Association of Fire Fighters, Washington, DC

Richard D. Gerkin, Jr., MD
Medical Director, Phoenix Fire Department Health Center; Associate Director, Samaritan Regional Poison Center, Good Samaritan Regional Medical Center, Phoenix, Arizona

Richard Gist, PhD
Director, Social Sciences and Social Services, Johnson County Community College, Overland Park, Kansas; Consulting Community Psychologist, Kansas City (Missouri) Fire Department, Kansas City (Missouri) Health Department, South Metropolitan Fire Protection District, Cass County, Missouri

Anne L. Golden, PhD
Assistant Professor, Division of Environmental and Occupational Medicine, Department of Community Medicine, Mount Sinai School of Medicine, New York, New York

Marella L. Hanumadass, MD
Assistant Professor of Surgery, Department of Surgery, University of Illinois College of Medicine; Chairman of the Division of Burn Services, Director of the Sumner L. Koch Burn Center, Cook County Hospital, Chicago, Illinois

Robert Harrison, MD, MPH
Occupational Health Branch, California Department of Health Services, Berkeley, California

Seth Michael Krosner, MD, FACS
Assistant Professor of Surgery, Department of Trauma and Burns, University of Illinois; Medical Director of the Burn Intensive Care Unit, Cook County Hospital, Chicago, Illinois

Philip J. Landrigan, MD, MSc
Division of Environmental and Occupational Medicine, Department of Community Medicine, Mount Sinai School of Medicine, New York, New York

Peter S. J. Lees, PhD, CIH
Department of Environmental Health Sciences, The Johns Hopkins University School of Hygiene and Public Health, Baltimore, Maryland

Jerrold D. Leikin, MD, FACP, FACEP
Associate Professor, Rush Medical College; Medical Director, Rush Poison Control, Chicago, Illinois

Steven B. Markowitz, MD
Associate Professor, Division of Environmental and Occupational Medicine, Department of Community Medicine, Mount Sinai School of Medicine, New York, New York

Barbara L. Materna, PhD, CIH
Occupational Health Branch, California Department of Health Services, Berkeley, California

Melissa A. McDiarmid, MD, MPH
Adjunct Associate Professor of Environmental Health Sciences, The Johns Hopkins University School of Hygiene and Public Health, Baltimore, Maryland

James Melius, MD, DrPH
Medical Director, Center to Protect Workers' Rights, Washington, DC

Peter Orris, MD, MPH, FACP, FACOEM
Associate Professor of Medicine, University of Illinois; Attending Physician, Division of Occupational Medicine, Cook County Hospital, Chicago, Illinois

Paul A. Reichelt, PhD
Associate Professor, Department of Public Health, Mental Health, and Administrative Nursing, College of Nursing, University of Illinois at Chicago, Chicago, Illinois

Nathaniel Rothman, MD, MPH
Department of Environmental Health Sciences, The Johns Hopkins University School of Hygiene and Public Health, Baltimore, Maryland

Cornelius H. Scannell, MD, MPH
Assistant Professor of Medicine, Division of Occupational and Environmental Medicine and Pulmonary Medicine, University of California at San Francisco, San Francisco, California

Randy L. Tubbs, PhD
Psychoacoustician, Division of Surveillance, Hazard Evaluations, and Field Studies; National Institute for Occupational Safety and Health, Cincinnati, Ohio

Virginia M. Weaver, MD, MPH
Instructor, Division of Occupational Health, Department of Environmental Health Sciences, The Johns Hopkins University School of Hygiene and Public Health, Baltimore, Maryland

Steven Joseph Woodall, MEd
Department of Fire Science, Eastern Arizona College, Payson, Arizona; Certified Associate Counselor, St. Luke's Employee Assistance Program, Phoenix, Arizona

PUBLISHED ISSUES
(*available from the publisher*)

April 1987	**Occupational Pulmonary Disease** Linda Rosenstock, MD, MPH, Editor
October 1987	**Workers with Multiple Chemical Sensitivities** Mark R. Cullen, MD, Editor
April 1988	**Worker Fitness and Risk Evaluation** Jay S. Himmelstein, MD, MPH, and Glenn S. Pransky, MD, MPH, Editors
July 1988	**The Petroleum Industry** Neill K. Weaver, MD, Editor
October 1988	**Psychiatric Injury in the Workplace** Robert C. Larsen, MD, and Jean S. Felton, MD, Editors
January 1989	**The Management Perspective** L. Fleming Fallon, Jr., MD, MPH, MBA, O.B. Dickerson, MD, MPH, and Paul W. Brandt-Rauf, ScD, MD, DrPH, Editors
April 1989	**Alcoholism and Chemical Dependency** Curtis Wright, MD, MPH, Editor
October 1989	**Problem Buildings** James E. Cone, MD, MPH, and Michael Hodgson, MD, MPH, Editors
January 1990	**Hazardous Waste Workers** Michael Gochfeld, MD, PhD, and Elissa Ann Favata, MD, Editors
April 1990	**Shiftwork** Allene J. Scott, MD, MPH, Editor
July 1990	**Medical Surveillance in the Workplace** David Rempel, MD, MPH, Editor
October 1990	**Worksite Health Promotion** Michael E. Scofield, PhD, Editor
January 1991	**Prevention of Pulmonary Disease in the Workplace** Philip Harber, MD, MPH, and John R. Balmes, MD, Editors
April 1991	**The Biotechnology Industry** Alan M. Ducatman, MD, and Daniel F. Liberman, PhD, Editors
July 1991	**Health Hazards of Farming** D. H. Cordes, MD, MPH, and Dorothy Foster Rea, MS, Editors
October 1991	**The Nuclear Energy Industry** Gregg S. Wilkinson, PhD, Editor
January 1992	**Back School Programs** Lynne A. White, Editor
April 1992	**Occupational Lung Disease** William S. Beckett, MD, and Rebecca Bascom, MD, Editors
July 1992	**Unusual Occupational Diseases** Dennis J. Shusterman, MD, MPH, and Paul D. Blanc, MD, MSPH, Editors
October 1992	**Ergonomics** J. Steven Moore, MD, and Arun Garg, PhD, Editors

1995 ISSUES

Effects of the Indoor Environment on Health
Edited by James M. Seltzer, MD
University of California School of Medicine
San Diego, California

Construction Safety and Health
Edited by Knut Ringen, DrPH,
Laura Welch, MD, James L. Weeks, ScD, CIH,
and Jane L. Seegal, MS
Washington, DC
and Anders Englund, MD
Solna, Sweden

Occupational Hearing Loss
Edited by Thais C. Morata, PhD,
and Derek E. Dunn, PhD
National Institute for Occupational
 Safety and Health
Cincinnati, Ohio

Firefighters
Edited by Peter Orris, MD
Cook County Hospital
Chicago, Illinois
and Richard M. Duffy, MS
International Association of Fire Fighters
Washington, DC
and James Melius, MD, DrPH
Center to Protect Workers' Rights
Washington, DC

1994 ISSUES

Occupational Skin Disease
Edited by James R. Nethercott, MD
University of Maryland
Baltimore, Maryland

Occupational Safety and Health Training
Edited by Michael J. Colligan, PhD
National Institute for Occupational
  Safety and Health
Cincinnati, Ohio

Reproductive Hazards
Edited by Ellen B. Gold, PhD, B. L. Lasley,
PhD, and Marc B. Schenker, MD, MPH
University of California
Davis, California

Tuberculosis in the Workplace
Edited by Steven Markowitz, MD
Mount Sinai School of Medicine
New York, New York

1993 ISSUES

The Mining Industry
Edited by Daniel E. Banks, MD
West Virginia University
Morgantown, West Virginia

De Novo Toxicants
Edited by Dennis J. Shusterman, MD, MPH
California EPA, Berkeley, California
and Jack E. Peterson, PhD, CIH
Alpine, California

Spirometry
Edited by Ellen A. Eisen, ScD
University of Massachusetts
Lowell, Massachusetts

Women Workers
Edited by Dana M. Headapohl, MD
St. Patrick's Hospital
Missoula, Montana

Ordering Information:
Subscriptions for full year and single issues are available from the publishers—
Hanley & Belfus, Inc., 210 South 13th Street, Philadelphia, PA 19107
Telephone (215) 546–7293; (800) 962–1892. Fax (215) 790–9330.

PREFACE

Fire and its control have been a central issue of human society since its inception. The ability of civilizations over the centuries to harness the productive power of combustion has often been a mark of their strength and power. The attention to prevention of destruction and death related to uncontrolled combustion has often been an index of the importance a government places on its common people. In this way, too, fire prevention has been a marker of a country's progress toward democracy and an adequate standard of living.

Yet, despite the significance of fire for society, the health and safety of those assigned to controlling it—firefighters—have frequently been neglected issues. It has been assumed that the hazards of firefighting are obvious and uncontrollable. Attention has frequently been focused on pension issues for disabled firefighters, and little time has been spent analyzing both the acute and chronic pathologies related to firefighting and how to prevent them.

This has begun to change during the last half century in many countries throughout the world. This increased attention to firefighter health and safety often has been stimulated by the firefighters themselves. In the United States and Canada, the International Association of Fire Fighters, AFL-CIO, the union representing most professional firefighters, has an impressive history of interest in the development of preventive programs to eliminate unnecessary hazards of firefighting. Starting in the 1960s, the union has organized a biennial conference on occupational safety and health in the fire service. This meeting now attracts more than 1,000 participants. The union also has organized a committee of medical advisors to provide technical assistance on health matters. With a broad concern for society, the IAFF has provided funding for residency training in occupational health at Johns Hopkins School of Hygiene and Public Health.

A fire service consensus organization, the National Fire Protection Association, is the oldest and most established group to promote health and safety in the fire services. The NFPA develops consensus regarding technical specifications for protective clothing, fire trucks, hoses, and all other aspects of fire equipment. Besides the equipment specifications, the NFPA provides consensus standards and guidance for local departments to provide occupational health programs and fitness standards for firefighters.

Fire-related deaths, injuries, and chronic diseases are often due to a combination of external stresses caused by fires and underlying subclinical incapacities of firefighters. These subclinical incapacities may predispose the firefighter to injury or death when confronted by the extraordinary physical demands of combating fires. Medical surveillance of firefighters needs to be directed toward identifying those subclinical incapacities prior to the time that they become a direct threat to the firefighter or his or her colleagues. With prior knowledge of these conditions, a firefighter may take steps to improve their functioning and reverse the development of a clinically significant disability.

The NFPA Standard 1582 (1992) recommends yearly medical certification of a firefighter's ability to perform his or her duties. The medical surveillance program is aimed at identifying conditions listed in chapter 3 of the standard that are either type A (disqualifying) or type B (disqualifying based on severity). The standard requires that the fire department arrange that this evaluation be conducted by a qualified physician of medicine or osteopathy, with expertise in occupational safety and health for emergency services.

The NFPA recommends that each firefighter evaluation include an interval medical and exposure history, height, weight, and blood pressure. The physical examination must look at vital signs; ear, nose, throat, and mouth; and the dermatologic, cardiovascular, respiratory, gastroenterologic, genitourinary, endocrine, musculoskeletal, and neurologic systems. Laboratory tests are required for audiometry, visual acuity, peripheral vision, and pulmonary function. Other testing is indicated if ordered by the physician. A medical examination is required every three years for those younger than 30, biannually for those in their 30s, and every year after 40.

Firefighters are required to exert maximal effort in emergency situations. These incidents are unpredictable and require immediate action and attention. Between these episodes the firefighter is frequently sedentary for periods varying from a few minutes to weeks depending on the community. General fitness, pulmonary function, and cardiac status are clearly the significant physiologic parameters needing careful attention during these surveillance evaluations.

Spirometry, with and without administration of bronchodilator medication, is an inexpensive, easily performed screening examination to assess subclinical pulmonary dysfunction. These tests are in wide use and enable a physician to identify asymptomatic individuals beginning to manifest decreased pulmonary function or reactive airway disease. This group of firefighters is then identified to undergo more elaborate pulmonary function evaluations. They are certainly an appropriate compromise with the more expensive evaluative techniques.

Due to the cost and questionable efficacy in asymptomatic adults, exercise stress testing has been an area of some controversy and is not currently recommended routinely. The Preventive Services Task Force of the Department of Health and Human Services noted in 1989 that exercise electrocardiographic testing was both a more sensitive and specific test than resting electrocardiograms. This testing provides an objective measure both of cardiac status and general physical fitness. It can detect subclinical cardiac disease as well as provide an index of decreasing fitness. The addition of contemporaneous pulmonary assessment and cardiac imaging techniques have improved and raised the cost of this modality still further. Currently these tests are in use only as part of a post myocardial infarction rehabilitation return to work assessment.

Other specific testing for infectious diseases and other specialized situations are covered in the relevant chapters of this text.

In addition to a physician knowledgeable in the health hazards of emergency services, a fire department today should expect to contract for a comprehensive occupational health program at no additional cost. These programs are based on the premise that injury care and preplacement physicals are important but alone do not constitute a complete program that emphasizes prevention in addition to treatment. A comprehensive occupational health program should include treatment of acute injuries, preplacement physicals, OSHA mandated exams (asbestos, lead, respirator, etc.), drug screening, referral and case management of workplace injuries, ergonomic evaluations, drug testing, substance abuse training and treatment, toxicology consultation, workplace evaluations, education through back schools and ergonomic programs, wellness programs, employee assistance programs, and rehabilitation services. The occupational health program staff should be familiar with all services, and they should meet the requirements and/or standards imposed by federal, state and local agencies such as Department of Transportation, Occupational Safety and Health Administration, and the Americans with Disabilities Act.

The editors hope that this volume will provide a comprehensive background for physicians and other health professionals involved in providing health care services to firefighters. Each individual chapter reviews an aspect of the health hazards of firefighting and their effect on firefighters. No initial review of this sort will cover all areas, yet it is hoped that this review will place enough information in one location to serve as a basic text for clinicians wishing to practice in this area as well as for departments and local unions wishing to arrange for this coverage.

Peter Orris, MD, MPH
James Melius, MD, DrPH
Richard Duffy, MS
GUEST EDITORS

PETER S. J. LEES, PhD, CIH

COMBUSTION PRODUCTS AND OTHER FIREFIGHTER EXPOSURES

From the Department of
  Environmental Health Sciences
The Johns Hopkins University
  School of Hygiene and Public
  Health
Baltimore, Maryland

Reprint requests to:
Peter S. J. Lees, PhD, CIH
Department of Environmental
  Health Sciences
The Johns Hopkins University
  School of Hygiene and Public
  Health
615 North Wolfe Street
Baltimore, MD 21205

The work environment of firefighters is unlike that of any other occupation. Firefighters are routinely exposed under harsh and uncontrolled conditions to extremely high and often lethal concentrations of an extremely wide array of chemicals. To exacerbate the problem, firefighters are frequently confronted with situations in which they have little or no knowledge of what chemicals are present. Unlike typical industrial occupational exposures, firefighters may be exposed to completely different chemicals from day to day or even from fire to fire. All of this takes place in situations where no means of exposure control beyond the self-contained breathing apparatus is feasible.

Products of combustion constitute the major source of firefighters' chemical exposures. Combustion is a complex chemical/physical process in which the materials present and the conditions under which they are combusted greatly affect the chemical products. The composition and concentrations of the resulting products are largely a function of the amount of oxygen present, i.e., ventilation and the thermal energy transferred to the burning material.[20] Combustion of a natural product such as wood is a relatively simple process, theoretically yielding only carbon dioxide and water in the presence of sufficient oxygen. Even a simple working fire involving only natural materials, however, does not represent this theoretical ideal; other more toxic chemicals are always produced. Combustion of synthetic products is even more complex. The combustion of the common plastic polyvinyl chloride (PVC) has been estimated to produce dozens of different chemicals.[41] In addition, flame retardants, fillers, plasticizers, ultraviolet light absorbers, antioxidants, lubricants, reinforcing fibers, peroxides, coupling agents, halogen stabilizers, biologic preservatives, anti-static agents, flow controls, coloring agents and other exotic chemicals are added to the PVC, further complicating the emissions inventory.[1] There is evidence that the combustion products of at least some of the additives, fire retardants in particular, are more toxic than those of the parent material.[34] Perhaps the only bright note is that oxygen depletion as a result of combustion is not thought to represent a significant hazard in a typical residential working fire.[16]

Highly monitored burns in various test facilities including full-scale houses and other structures have shown that "since so many variables such as ignition source, materials, ventilation, etc., are involved in any real fire, it is difficult to accurately predict the composition of a fire atmosphere at any given time or place."[23] A generalized characterization of the fire environment and firefighter exposures is possible, however, because a fairly basic set of simple products of combustion is associated with nearly all fires.

Partial combustion products constitute the major type of exposure. Historically firefighters confronted with a dwelling fire or even an industrial fire were exposed to the combustion products of natural materials. Even today, with the vast majority of fires occurring in residences and with bedding and upholstered furniture cited as the source of almost half of all such fires,[5] natural materials constitute the major source of fuel. The combustion products of such fires are relatively simple and have relatively low toxicity. These natural products in the home and other structures have been supplemented since the mid 20th century by a growing array of synthetic products, predominantly plastics and organic chemicals. These chemicals are increasingly incorporated in carpets, furnishings, wall coverings, and even structural elements in homes and are a major part of commercial buildings. Industrial fires, once fairly similar to residential fires, are increasingly complex events in which a bewildering array of chemicals may be released during combustion.

Even firefighters involved in forest and grasslands fires are no longer exposed to the simple products of wood and grass combustion; they also are exposed to herbicides[32] and aerial fire retardant chemicals.

In addition to combustion products, firefighters may be exposed to the very tools they use for suppressing fires; for instance, Halon[25] and (formerly) carbon tetrachloride[13] are themselves toxic. Even diesel exhausts from fire apparatus constitutes another source of potentially toxic exposure.[15]

Over the last two decades, hazardous materials response has greatly expanded the number of chemicals to which firefighters are potentially exposed. While the number of chemicals released in a given incident presents a simpler exposure situation than during combustion, firefighters may be exposed to virtually all of the tens of thousands of chemicals used in commerce during HazMat incidents.

SOURCES OF DATA

Despite the striking and often dramatic nature of these exposures, characterization of the exposures of firefighters makes up a strikingly small body of literature. Much of this paucity undoubtedly derives from the unavailability until recent times of sampling and analytic methods and equipment compatible with the fire environment along with concern for firefighter safety during sample collection. Sampling methods devised for low-level industrial exposures are often overloaded in the fire environment. The complex nature of combustion products present may interfere with standard methods. Pragmatically, it is impossible to schedule sampling; incidents are unplanned and exposures are typically brief. As such, virtually all direct measures of firefighter exposure come from common residential fires; usual and less common exposures from industrial fires remain unmeasured.

Estimates of firefighter chemical exposures may come from direct measures or may be inferred from indirect or surrogate measures. Personal breathing zone measurements represent the best source of direct exposure data, although area measures in actual working fires also may qualify. These are obviously the most reliable estimates of firefighter exposure. Given the wide variety of types of fires, combustion conditions, and firefighting techniques, a wide variety of concentrations can be expected, even for the most common substances. However, most of these measures are atmospheric measurements that do not represent actual inhalation concentrations; the use of self-contained breathing apparatus (SCBA) will substantially reduce exposures if worn consistently.[9,27]

Indirect measures of exposure include medical and epidemiologic literature and emissions measurements from test fires. Epidemiologic studies of firefighter mortality do not include reliable measures of exposure. As such, epidemiologic results showing an excess of some disease can only be used to generate hypotheses inferring high exposure to some chemical.[12,17,22,23] Much of the early literature concerning firefighter exposures is derived from clinical studies and case reports of firefighters and fire victims. In these studies, the firefighter acts as the environmental sampler, and the biologic endpoint, e.g., carboxyhemoglobin, is measured.[35,38] Given the available technology, it is not surprising that most of these studies focus on carbon monoxide (as measured by carboxyhemoglobin) and smoke (as measured by pulmonary function tests). Recent studies have begun to relate exposure to organic chemical combustion products to other biologic markers, such as DNA adducts.[37] Exposure in these studies is inferred from the biologic measures; in some cases it is possible to back-calculate an environmental exposure concentration.

Another indirect source of exposure information is measurement of the products of combustion under controlled conditions. While these studies have produced a bewilderingly long list of chemicals, their direct applicability to actual fire conditions is not known. Given the great impact of combustion conditions on the types and concentrations of products produced, exposures identified should be viewed as surrogate measures that represent only possible exposures. There is a small body of literature in which residues from actual fires, including water, are analyzed to infer exposures. How the results of these studies relate to actual exposures is also unknown.

Assessment of firefighter exposures is a complex problem. This chapter summarize in detail the literature on direct measures of exposure and briefly enumerates other data. Direct measures are the most reliable estimates of firefighter exposure. The literature focuses on the main constituent chemicals in the fire environment, although probably not the most toxic. Studies conducted at Harvard in the early 1970s involving 72 and 190 fires,[9,10,16,42] studies conducted by Columbia University in Buffalo, N.Y., involving 14 fires,[7] and studies conducted by the National Institute for Occupational Safety and Health to field test the efficacy of firefighter SCBA involving 22 fires[24] form the heart of these data. More recent studies have begun to examine exposures of persons fighting forest fires and wildfires.[30,36]

Most of the studies presenting direct measures of firefighter exposure appropriately focus on the extreme high end of the exposure distribution. The range of exposures is presented in some cases and the entire distribution in others. Some studies present average exposures, although this measure may be misleading because the acute health effect/immediate danger is associated with the high concentrations. Longer-term average exposure measures, however, may be more useful in assessing the risk from exposure to carcinogenic chemicals. None of these studies adequately present the time period over which exposure measures were collected, but the technical report for one study presents this information for each sample.[10] Time is a crucial factor in assessing exposure and possible health effects. In studies of residential fires, exposure measurements usually last only minutes. In forest and wildfire studies, they are usually several hours.

Table 1 presents some guidance to help evaluate the magnitude of exposures reported in the literature. The permissible exposure limits (PEL) promulgated by the Occupational Safety and Health Administration, the recommended exposure limits (REL) published by NIOSH, and the threshold limit values (TLV) adopted by the American Conference of Governmental Industrial Hygienists represent allowable/recommended 8-hour time-weighted average concentrations and, as such, are more directly useful for industrial settings. They were designed for the most part to minimize risk over a 45-year working lifetime and presuppose exposure for the entire period. Comparison of firefighter exposures to these standards may be somewhat misleading. The short-term exposure limit (STEL) is a shorter term, 15-minute, average concentration, designed in part to prevent acute effects of exposure. The immediately dangerous to life and health (IDLH) data represent the concentration from which a person could escape within 30 minutes without irreversible health effects or symptoms that would interfere with escape. The short-term lethal concentration (STLC) is a concentration that may be fatal within 10 minutes. These short-term measures are designed to prevent acute effects, often unconsciousness, or death. They do not address the question of more subtle long-term effects of exposure at lower concentrations represented by PEL and TLV. Exposure concentrations presented in the literature are measurements of contaminant concentrations in the air and do not necessarily represent actual firefighter exposure. In part due to some of the findings of early studies, firefighters now make extensive use of SCBA. Using the usually assigned protection factor of 2,000,[6] actual exposures may be theoretically calculated by dividing the concentrations presented by 2,000. This calculation assumes, of course, that the firefighter wears the SCBA during the entire exposure period. Limited measurements made inside the facemask of SCBA are presented in the NIOSH study.[24] In general, these measurements show a significant reduction in exposure, but far less than the 2000-fold decrease postulated above. This difference is undoubtedly due primarily to wearing the SCBA for varying amounts of time. One study has reported the 70% of firefighters remove their mask at some time during knockdown and that as many as a third of firefighters wear their mask less than half of the time.[24]

TABLE 1. Exposure Standards, Guidelines, and Guideposts Useful for the Interpretation of Firefighter Exposure Data

Chemical	Permissible Exposure Limit	Threshold Limit Value	Recommended Exposure Limit	Short-term Exposure Limit	Immediately Dangerous to Life and Health	Short-term Lethal Concentration
Acetaldehyde	200	—	LF	25(C)[1]	2,000	—
Acrolein	0.1	0.1	0.1	0.3[1]	2	30–100
Benzene	1	10	0.1	5[2]	500	20,000
Carbon dioxide	5,000	5,000	5,000	30,000[1]	40,000	100,000
Carbon monoxide	50	25	35	200[2]	1,200	5,000
Formaldehyde	0.75	—	0.016	0.3[1]	20	—
Hydrogen chloride	0.5	—	0.5	5(C)[1]	50	>500
Hydrogen cyanide	10	—	—	4.7(C)[1]	50	350
Nitrogen dioxide	5(C)	3	—	5	20	>200
Sulfur dioxide	5	2	2	5	100	>500

All concentrations expressed as ppm.
1 = American Conference of Governmental Hygienists
2 = Occupational Safety and Health Administration
LF = lowest feasible concentration
(C) = ceiling not to be exceeded

EXPOSURE MEASUREMENTS FOR COMBUSTION PRODUCTS

Particulates and Smoke

Smoke is the most obvious product of combustion and by far the most complex. While smoke can be viewed simplistically as partially combusted carbonaceous material, it is in reality a bewildering array of organic and inorganic compounds original to the combusted material, decomposed from the original material, or formed from the combustion process. A different set of products and product concentrations exists for every material burned and every set of combustion conditions. Beyond the complications imposed by the combustion process, something apparently simple, like a wood-frame dwelling fire, is far more complex depending on what the wood was painted with, if it had a preservative, and if it had been treated recently for termites. Information on the specific chemical nature of smoke and the resulting exposures is treated on a chemical-by-chemical basis in the following sections. In this section, smoke is treated as a simple particulate aerosol. In addition to carbon particles, silica, fluoride, aluminum, lead, acids, bases, and phenols have been identified in the particulate phase of smoke.[26]

Despite many papers on the respiratory effects of smoke inhalation, personal breathing zone measures of the particulate concentration in working fires have rarely been made. The earliest personal breathing zone measures date from the Harvard studies.[9,10,16,42] The initial report, based on 20 samples, indicated a geometric mean particulate concentration of 21.5 mg/m^3 (GSD=4.7). Approximately 15% of all samples were in excess of 100 mg/m^3.[16] In subsequent studies involving 237 samples, and concentrations ranging up to 18,000 mg/m^3, the geometric mean concentration increased only slightly.[10,42]

An extensive study of firefighter exposures in Buffalo[7] included only five measures of total particulate. All were collected at five fires in which the smoke intensity was described as low or moderate. Concentrations ranged from 10.1–344.4 mg/m^3; the highest concentration was measured in a building fire with "low" smoke intensity and is a clear outlier; the next highest measured concentration was 38.3 mg/m^3.

NIOSH's study of 22 mostly residential fires determined that total particulate concentrations ranged from not detectable (ND)–560 mg/m^3 during the knockdown phase to ND–45 mg/m^3 during the overhaul phase.[24] Mass median particle diameters were 10 μm during knockdown and 1 μm during overhaul at one fire. Some of the particulate was nonasbestiform fibers; concentrations ranged from background (BG)–0.21 f/ml to BG–0.36 f/ml for knockdown and overhaul respectively.

Another NIOSH study of exposures associated with forest firefighting included six measures of respirable particulate exposures.[36] Respirable particulate was measured with the standard Dorr-Oliver cyclone. Samples were of longer duration than in studies of structural fires and ranged from 0.6–1.7 mg/m^3 (mean=1.2 mg/m^3).[36] Another more extensive study of particulate concentrations in California wildland fires showed respirable particulate exposure to range from 0.327–5.15 mg/m^3 and to average 1.75 mg/m^3 (n=22) during mop-up on the fireline and to range from 0.235–2.71 mg/m^3 and to average 1.15 mg/m^3 (n=20) during a prescribed burn.[30]

Another source of particulate smoke is the exhaust of diesel engines typically used in fire apparatus. In a set of 228 personal samples collected in firehouses, total airborne particulate concentrations were found to range from 30–480 μg/m^3. The authors predicted an average concentration of 300 μg/m^3, of which 225 μm/m^3 was attributable to the diesel exhaust and the remainder to background and to cigarette smoking. Real-time continuous measurements showed total particulate concentrations frequently to peak above 1 mg/m^3 (1,000 μg/m^3) during departure of fire apparatus.[15]

Carbon Dioxide

Carbon dioxide is a product of all combustion processes and, as a result, is present in all fire environments. Concentrations range from normal background (350–400 ppm) to in excess of 100,000 ppm in test fires.[10] Despite its acknowledged prevalence and known ability to act as a simple asphyxiant, although of relatively low toxicity, carbon dioxide exposures of firefighters have been measured in only a few studies.

In the earliest set of measurements in the Harvard studies, carbon dioxide was not detected in any of the 63 samples collected. Since the analytical method employed had a limit of detection of 0.26% (2,600 ppm), it is better stated that no concentration in excess of 2,600 ppm was detected.[16] Using a more sensitive analytic method, subsequent Harvard studies found that concentrations in 89 samples averaged 3,350 ppm. The highest observed concentration (70,000 ppm) exceeded the IDLH, and three measurements exceeded to STEL.[10,42]

Carbon dioxide concentrations measured in the more recent NIOSH studies agree closely with findings of the Harvard studies and provide additional information. During the knockdown phase, carbon dioxide concentrations ranged from 350–5,410 ppm in 19 samples and failed to show the extreme peaks seen in the Harvard studies. During overhaul, concentrations ranged from 130–1,420 ppm. Concentrations inside the SCBA mask ranged from 460–21,300 ppm, probably not reflecting the environment but rather the metabolic contribution from strenuous activity.

Carbon Monoxide

Carbon monoxide is present in virtually all fire environments. It is a product of the incomplete combustion that describes every fire, especially during the earliest phases and during fire suppression when temperatures have been lowered. Carbon monoxide is a highly toxic substance and has long been recognized as a critical exposure in the fire environment. Early studies ascribed virtually all of the toxic effects of fire to carbon monoxide, a claim now recognized as somewhat exaggerated. Nonetheless, carbon monoxide exposures are critical to assessing firefighter exposure. Many measurements have been made; technology for assessing long- and short-term average exposures have long been available. Recent studies take advantage of newer technology that permits real-time measurement with storage of concentration data to a data logger.

The earliest published measures of firefighter exposure to carbon monoxide are derived from a Harvard study designed to determine the protection factor required for firefighter respiratory protection devices.[9] Measurements made during 1,329 minutes of sampling under more severe than usual conditions showed carbon monoxide concentrations to exceed 500 ppm approximately 29% of the time. The highest measured instantaneous concentration was 27,000 ppm.[9] Measurements made a decade later in Los Angeles generally confirm these short-term exposure levels.[4] In this study, peak concentrations of 3,000 ppm were measured. Approximately 48% of the measurements were above 500 ppm.

Another early Harvard study used short-term measures, with an average duration of less than 10 minutes. As expected for environmental measures, concentrations were log-normally distributed with a geometric mean concentration of 110 ppm in approximately 65 samples. Concentrations ranged from less than 2 ppm to more than 1,000 ppm. Approximately 3% of all measurements exceeded 1,000 ppm.[16]

Treitman and Burgess[42] in subsequent studies found an even wider range of carbon monoxide exposures, 0-4,800 ppm in 110 samples. The average exposure was 320 ppm and the STEL of 400 ppm was exceeded in approximately 15% of the fires. Brandt-Rauf found a similar range of exposures in 14 fires. Carbon monoxide concentration ranged from 11.4–1087 ppm in 22 samples. Findings were similar to the Harvard study in that about 19% of the measurements exceeded the STEL of 400 ppm.[7]

The findings of the NIOSH study[24] further confirm earlier carbon monoxide findings. During the knockdown phase, carbon monoxide concentrations ranged from BG–1,900 ppm in 33 samples. Unlike previous studies, however, fully 33% of the measurements exceeded the 400 ppm STEL. Approximately 10% of the samples were above the 1,500 ppm IDLH. During the overhaul phase, concentrations were considerably lower, ranging from BG–82 ppm in 7 samples. Carbon monoxide measurements made inside firefighters' SCBA masks ranged from 1–105 ppm in 6 samples. No correlation is presented between specific environmental exposures and the corresponding concentration inside the mask. Real-time measurement of carbon monoxide concentrations at one fire showed rapidly changing conditions in which exposure varied from less than 10 ppm to more than 1,000 ppm over a period of less than 5 minutes.[24] Such variation is not represented in time-averaged measurements.

As expected because of the lack of confinement, studies of firefighter exposures during forest and wildfire suppression reveal strikingly lower exposures to carbon monoxide. The study of wildfires showed concentrations to range from 1.4–38 ppm in 46 samples during fireline mop-up and a prescribed burn. The average concentration was 14.4. ppm, and the average sampling period was 3.5 hours. Concentrations were higher during evening hours when inversion conditions were frequently present. The highest exposures, up to 300 ppm on an instantaneous basis, were associated with operators of gasoline-powered pumping engines.[30] Results measured using several different methods were comparable.[31] A somewhat smaller set of samples from crews fighting forests fires showed carbon monoxide concentrations to range from 1.2–24.2 ppm in 19 samples. The average concentration was 11.5 ppm, and the average sampling period was approximately 9 hours. Carbon monoxide concentrations corresponded to subjective estimates of smoke density.[36]

Environmental carbon monoxide concentrations have been back-calculated from carboxyhemoglobin concentrations. Using standard techniques, exposure of nonsmoking firefighters extinguishing Australian bushfires was estimated to average 17 ppm.[8] This estimate is in close agreement with the environmental measures reported above.

Nitrogen Dioxide

Nitrogen dioxide is formed through the degradation of nitrogen-containing materials in fabrics and cellulose nitrate or through the fixation of atmospheric nitrogen. Authors of the studies conducted at Harvard concluded that conditions in typical residential fires were not conducive for either of these means of formation.[10,42] In the preliminary Harvard study, nitrogen dioxide was detected in 8 of 90 samples collected. Detectable concentrations ranged from 0.02 – 0.89 ppm.[16] In subsequent studies at Harvard, nitrogen dioxide was detected with much greater frequency. Measured concentrations ranged from 0–9.5 ppm in 240 samples. Four samples exceeded the 5 ppm STEL, and the highest exposure measured over a 5-minute period was 8.3 ppm.[10,42]

In the only published measurement of nitrogen dioxide exposure of forest firefighters, concentrations were below the limit of detection in all ten samples collected.[36] The limit of detection was not presented, but sample times were long.

Hydrogen Chloride

Hydrogen chloride may be present in the fire environment as a gas or as an aerosol. It is formed from the decomposition of chlorine-containing compounds. In structural and automobile fires, plastics are the typical source of chlorine. Many plastics, including PVC, chlorinated acrylics, and certain flame retardant chemicals, enjoy extensive use.

Despite the greatly increased use of plastics, hydrogen chloride is detected relatively infrequently in studies of firefighter exposure. Gold was able to detect hydrogen chloride in only 5 of 90 samples. In each of the fires in which hydrogen chloride was detected, there was general involvement of a room and its contents; the five measured concentrations ranged from 18–150 ppm.[16] A follow-up study at Harvard detected hydrogen chloride more frequently, although it was still only found in 36% of the 242 measurements. Concentrations ranged from 0–280 ppm with an average concentration for fires in which hydrogen chloride was detected of 36 ppm. In 90% of the fires, hydrogen chloride concentrations were less than 50 ppm. Five samples exceeded the 100 ppm IDLH, and 73 samples exceeded the 5 ppm STEL.[10,42]

More recently, the Buffalo studies detected hydrogen chloride in only two of the 19 samples collected and analyzed. Concentrations of 2.17 and 13.3 ppm were detected.[7] Subsequent studies by NIOSH detected hydrogen chloride in only two samples from 22 fires. The two measured concentrations, less than 2 ppm (estimated from graph) and 8.5 ppm, were detected during the knockdown phase of fire suppression. Hydrogen chloride was not detected during overhaul.[24]

While a significant portion of this variability is due to variability in the fire environments measured, differences in the sampling and analytical measures may also account for a major part of the apparent difference. The colorimetric tubes used in the Buffalo studies[7] are more likely to lead to erroneous reading than the others. In one of the studies,[24] 99% of the hydrogen chloride was found on the fiberglass support material rather than the adsorbent collection media. This lead the investigators to conclude that most of the hydrogen chloride was present in the aerosol rather than the gaseous form.

Hydrogen Cyanide

Hydrogen cyanide is a particularly toxic gas derived primarily from the combustion of nitrogen-containing polymers such as nitriles, polyamides, nylons, and polyurethane. It also can be formed during the combustion of paper, silk, and wool. Studies of firefighter exposures have found hydrogen cyanide to be increasing in prevalence, perhaps indicating the increased use of the synthetic polymers listed above.

In the preliminary Harvard studies, hydrogen cyanide was detected in 43 (approximately half) of the samples. In the samples in which hydrogen cyanide was detected, concentrations ranged from 0.02 to nearly 5 ppm. For all samples, however, the geometric mean concentration was 0.04 ppm with a geometric standard deviation of 8.3, indicating large variability. Four of the six samples in which concentrations exceeded 1 ppm were associated with either mattress or vehicle fires.[16] In subsequent studies at Harvard, hydrogen cyanide was detected in only about 10% (27 of 253) of the fires measured. This difference was attributed by the authors to a wider spectrum of fires in the second (larger) study. The highest concentration measured in a sample of more than 5 minutes duration was 3.6 ppm, while a concentration of 6.1 ppm was measured in a shorter duration sample.[10,42]

Hydrogen cyanide was detected in 10 of 26 samples (8 of 14 fires) collected as a part of the Buffalo studies. Measured concentrations ranged from 0–75 ppm, although the second highest concentration was 10 ppm. (The authors acknowledge the possibility of erroneous measurements from interference with the colorimetric detector tube system used for measurement.) If real, this highest concentration was significantly above the IDLH. Four of the five highest concentrations were associated with "intolerable" or "high" smoke density. Only two fires with "intolerable" smoke did not have high hydrogen cyanide concentrations.

In measurements made in 22 fires, NIOSH detected hydrogen cyanide during the knockdown phase in 12 fires and during the overhaul phase in three fires. Concentrations ranged from ND-23 ppm and ND-0.4 ppm, respectively. Three samples exceeded the 10 ppm STEL in the knockdown phase. Approximately 20% of the samples exceeded 1 ppm during the knockdown but none did during overhaul.[24]

Sulfur Dioxide/Sulfuric Acid

Sulfur dioxide is formed during the combustion of sulfur-containing substances. Upon contact with water used for fire suppression, a significant amount may be converted to sulfuric acid. Neither of these substances was measured in the Harvard studies. In the Buffalo study investigators found detectable concentrations of sulfur dioxide in 12 of 26 samples from 14 fires. Concentrations ranged from 0–41.7 ppm, although the second highest concentration was 2.5 ppm. Interestingly, this singular high sample was collected from the same firefighter as the singularly high hydrogen cyanide sample discussed above.[7] Sulfuric acid concentrations were measured in the NIOSH study. During knockdown, sulfuric acid was detected in three of 22 fires with concentrations reported to range from ND-8.5 mg/m^3. During overhaul, sulfuric acid was detected in two fires with concentrations reported to range from ND-0.9 mg/m^3.[24]

In Reh's measurements of sulfur dioxide exposure among 11 forest firefighters, concentrations ranged from 0.2–2.9 ppm, and the average was 1.5 ppm.[36]

Acrolein

Combustion of wood, cotton, paper, plastics including styrene and polyolefins, and oils and fats containing glycerol may result in the production of acrolein, a highly irritant gas that can cause death in a relatively short time at low concentrations. Chamber studies have showed acrolein evolution rates to vary by a factor of over 35x depending on the material that is burned.[41] Initially detected in the Harvard study, it has repeatedly been shown at high concentrations in other studies. Burgess and Treitman reported that while almost 40% of the samples showed concentrations less than 0.1 ppm, 66 of the 118 samples (56%) were above the STEL and five (4%) were above the IDLH of 5 ppm. The single highest sample was 98 ppm, a rapidly fatal exposure for an unprotected firefighter.[10,42]

Nearly 20 years later, these findings were duplicated in the NIOSH study. During knockdown, measured concentrations ranged from ND-3.2 ppm. About half of the exposures exceeded the 0.3 ppm STEL. During knockdown, concentrations ranged from ND-0.2 ppm, with acrolein being detected in only one of the 22 fires. Only one of the samples collected inside the SCBA mask yielded a detectable concentration, 0.9 ppm.[24]

Studies of forest and wildfire exposures have yielded lower concentrations. The single sample analyzed for acrolein in wildfire studies showed a concentration of 0.52 mg/m^3 (0.23 ppm).[30] In studies of forest fire exposures, the five samples showed concentrations from ND-0.01 ppm.[36]

Formaldehyde and Acetaldehyde

Other somewhat less toxic aldehydes, including formaldehyde and acetaldehyde, are formed in a manner similar to the more toxic acrolein. These substances have only been measured in the more recent studies as the result of improvements in sample collection technology. In the Buffalo study, formaldehyde was detected in six of the 24 samples. Concentrations ranged from 0.1–8.3 ppm, with the second highest concentration of 3.3. ppm.[7] Virtually identical results are presented in the NIOSH study, in which concentrations during knockdown were reported to range from ND–8 ppm in 22 samples, with formaldehyde detected in 73% of the samples. The two highest samples were in excess of the STEL. During overhaul, concentrations were lower, ranging from ND–0.4 ppm, with formaldehyde detected in 23% of the samples. Slightly lower concentrations were detected inside masks; ND–03 ppm, with formaldehyde detected in 23% of the samples. Acetaldehyde concentrations followed a similar pattern: ND–8.1 ppm during knockdown with 95% positive samples; ND–1.6 ppm during overhaul with 23% positive samples; and ND–0.9 ppm inside the masks with 18% positive samples.

As expected, limited study of aldehyde exposures related to forest and wildland fires indicates lower exposures. Formaldehyde was detected in all 30 samples collected during the study of wildland fires. Concentrations ranged from 0.048–0.42 mg/m^3 (0.04–0.34 ppm); the mean was 0.16 mg/m^3 (0.13 ppm).[30] A smaller study by NIOSH also detected formaldehyde in each of the five samples collected. Concentrations ranged from 0.02–0.07 ppm; the mean was 0.05 ppm.[36] Results of acetaldehyde sampling present a similar picture. During the study of wildland fires, it was detected at low concentrations in 80% of the samples collected. Concentrations ranged from ND–0.15 mg/m^3 (ND–0.08 ppm); the mean was 0.078 mg/m^3 (0.04 ppm) (n=24).[30] In the NIOSH study acetaldehyde was detected in all of the five samples. Concentrations ranged from 0.01–0.04 ppm; the mean was 0.02 ppm.[36]

Benzene

A staggering number of organic chemicals are produced by the combustion, partial combustion, pyrolysis, and/or thermal degradation of the natural (mostly wood) and synthetic (mostly plastic) products involved in residential and forest/wildland fires. Virtually every class of organic chemicals can be produced under some conditions. Benzene frequently has been measured as a simple surrogate for the wide array of organic chemicals because of its basic nature, the relative ease of sampling and analysis, and because of its combined acute and long-term toxic effects.

Not surprisingly, benzene is detected in nearly all fires. In the Harvard study, it was detected in 92% of the 197 samples. Measured concentrations ranged from ND–164.9 ppm. The second highest sample was 71.5 ppm; in all, 33 samples (17%) exceeded the STEL.[10,42]

More recent studies have generally confirmed these findings. Benzene was detected at about the same frequently but at higher concentrations in the Buffalo studies. It was detected in 69% of the 25 samples with the concentrations ranging from ND–250 ppm, with an average of 56 ppm. In all instances in which benzene was detected, it was present in concentrations above 5 ppm; 19% of all samples were above 100 ppm.[7] The NIOSH studies showed somewhat lower concentrations and a lower frequency of detection. Of particular note, no differences were observed in the concentrations measured during knockdown and in the face masks. Measured concentrations during knockdown ranged from ND–22 ppm; during overhaul they ranged from ND–0.3 ppm; and concentrations in the mask ranged from ND–21 ppm.[24]

Other Volatile Organic Chemicals

While there has been extensive characterization of other organic products of combustion, actual firefighter exposure measurements are rare and only recent. While universally present at concentrations far below levels at which acute effects would be observed, many of these substances are known or suspected carcinogens. Qualitative exposure information may be derived from clinical studies of the organic compounds present in firefighters' blood. For example, in a study of persons fighting the Kuwait oil fires, benzene, ethylbenzene, m-/p-xylene, o-xylene, styrene, and toluene were identified in the blood.[14]

A limited number of organic substances were measured during the Buffalo study. Maximum measured concentrations were as follows: dichlorofluoromethane 12.1 ppm, methylene chloride 0.278 ppm, trichloroethylene 0.181 ppm, chloroform 1.92 ppm, perchloroethylene 0.138 ppm, toluene 0.275 ppm, and trichlorophenol 0.129 ppm.[7]

As a part of the NIOSH study, a number of polynuclear aromatic (PNA) combustion products were quantified during knockdown and overhaul. PNAs were always detected in the knockdown samples, but were far less prevalent in the overhaul samples. Maximum measured concentrations were as follows: acenaphthene 100 $\mu g/m^3$, phenanthrene 100 $\mu g/m^3$, anthracene 30 $\mu g/m^3$, fluoranthene 60 $\mu g/m^3$, pyrene 70 $\mu g/m^3$, benz(b)anthracene 30 $\mu g/m^3$, chrysene 20 $\mu m/m^3$, benzo(b)fluoranthene 12 $\mu g/m^3$, benzo(k)fluoranthene 6 $\mu g/m^3$, benzo(e)pyrene 40 $\mu g/m^3$, benzo(a)pyrene 20 $\mu g/m^3$, indenol(1,2,3-cd)pyrene 20 $\mu g/m^3$, dibenz(a,h)anthracene 5 $\mu g/m^3$, and benzo(ghi)perylene 10 $\mu g/m^3$.[24] Additional substances were identified but not quantified in this and other studies (Table 2).

Exposures to most of the PNAs enumerated above have been quantified in forest and wildfires. Exposures are typically three to four orders of magnitude lower than reported above.[30,36]

TABLE 2. Chemicals with Qualitative Firefighter Exposure Information

Substance	Atlas et al.[3]	Henricks et al.[21]	Lowry et al.[29]
Acetic acid		X	
Acetic acid, methylester			X
Acetophenones		X	
Acrylic acid		X	
Aliphatic acids (C5-C14)		X	
Aliphatic alcohols (C2-C4)		X	
Aliphatic aldehydes (C6-C10)		X	
Aliphatic amines (C1-C2)		X	
Aliphatic hydrocarbons (C4-C7)		X	
Aliphatic ketones (C3-C5)		X	
Aliphatic ketones (C6-C8)		X	
Alkylbenzenes (C6-C8)		X	
Alkylbenzenes (C9-C10)		X	
Alkylphenols (C8-C9)		X	
Ammonia		X	
Anthracene	X		
Aromatic amines		X	
Benzaldehyde		X	
Benzofluoranthene	X		
Benzofurans		X	
Benzoic acid		X	
Benzo(ghi)perylene	X		
Benzo(e)pyrene	X		
Benzo(a)pyrene	X		
Bisphenol-A		X	
1,3-Butadiene			X
Butane			X
1-Butanol			X
2-Butanone			X
2-Butene			X
3-Butene			X
Butonic acid			X
Butylmethacrylate		X	

(Continued on page 702)

TABLE 2. Chemicals with Qualitative Firefighter Exposure Information *(Continued)*

Substance	Atlas et al.[3]	Henricks et al.[21]	Lowry et al.[29]
Butyraldehydes		X	
Butyric acid		X	
Chloromethane			X
Chrysene	X		
Cresols		X	
Crotonaldehyde		X	
Cyclohexadiene			X
1,4-Cyclohexadiene			X
Cyclohexane			X
1,3-Cyclopentadiene			X
Cyclopentane			X
Dibenzofluoranthene	X		
Dichlorodifluoromethane			X
Dichloromethane		X	X
1,3-Dimethyl benzene			X
2,3-Dimethyl-1-butene			X
1,1-Dimethyl cyclopentane			X
1,3-Dimethyl cyclopentane			X
1,2-Dimethyl cyclopentane			X
1,2-Dimethyl cyclopropane			X
2,5-Dimethyl furan			X
2,4-Dimethyl heptane			X
2,2-Dimethyl hexane			X
2,4-Dimethyl hexane			X
2,3-Dimethyl hexane			X
2,5-Dimethyl hexane			X
Dimethyl-methylane-bicyclo-heptane			X
2,3-Dimethylpentane			X
Ethanol			X
Ethyl benzene			X
Ethyl cyclopentane			X
3-Ethyl pentane			X
Formic acid		X	
Furan			X
Fluoranthene	X		
Heptane			X
1-Heptane			X
Hexane			X
Hexene			X
2-Hexene			X
1-Isocyano butane			X
1-Isocyanato-2-methyl propane			X
Methacrolein		X	
Methacrylic acid		X	
Methanol		X	
2-Methoxy-2-methyl propane			X
2-Methyl butane			X
2-Methyl-1-butene			X
3-Methyl-3-buten-2-one			X
Methyl cyclohexane			X
Methyl cyclopentane			X
2-Methyl furan			X
2-Methyl-2-hepten-4-one			X
3-Methylhexane			X

(Continued on page 703)

TABLE 2. Chemicals with Qualitative Firefighter Exposure Information *(Continued)*

Substance	Atlas et al.[3]	Henricks et al.[21]	Lowry et al.[29]
2-Methylhexane			X
Methylmethacrylate		X	
3-Methylpentane			X
2-Methylpentane			X
2-Methylpentene			X
2-Methyl-2-propanol			X
2-Methyl-1-propene			X
Nitrosamines		X	
1,1′-Oxybis-ethane			X
PAH-alcohols		X	
1,4-Pentadiene			X
1,3-Pentadiene			X
1,2-Pentadiene			X
Pentane			X
2-Pentene			X
3-Penten-2-one			X
Perylene	X		
Phenanthrene	X		
Phenol		X	
Phthalic anhydride		X	
Polyaromatic hydrocarbons (PAH)		X	
Propane			X
2-Propanol			X
2-Propanone			X
1-Propene			X
Proprionaldehyde		X	
Proprionic acid		X	
Pyrene	X		
Salicylaldehyde		X	
Tetrahydro-2-methyl furan			X
Toluene			X
Trichlorofluoromethane			X
Trichloromethane			X
Trimethyl-bicyclo heptane			X
2,3,5-Trimethyl hexane			X
2,2,4-Trimethyl pentane			X
Trimethylphenylindane		X	
Triphenylene	X		

Other Chemicals

Several other potential exposures of firefighters are of particular note, largely because of their potential toxicity. Unfortunately, exposure to these chemicals is largely inferred through indirect means. Literature shows biologic evidence of exposure and sometimes severe health effects in firefighters following fires with herbicides and pesticides,[32,39,43] diisocyantes,[11] and PCBs and related dibenzodioxins and dibenzofurans,[19,40] to name a few. Most of these exposures occurred as a result of industrial fires, and because they are so rare it is unlikely that they can ever be studied as extensively as the products of residential fires.

EXPOSURE MEASUREMENTS FOR HAZARDOUS MATERIALS RESPONSE

Within the last 20 years, there has been a greater and more organized response by firefighters to actual or potential spills and other releases of hazardous materials. As such, firefighters are potentially exposed to any and all of the tens of thousands of chemicals used in daily commerce. To date, a comprehensive study of exposures associated with HazMat response has not been published. There are a few scattered reports in the literature concerning exposure to specific substances at specific incidents. There is, however, a potentially large and untapped database of exposure data available from onsite sampling done by increasingly sophisticated HazMat teams.

Even more than is the case for firefighting, simple measurements of concentrations of contaminants in air should not be equated with HazMat firefighter exposure. Typical HazMat response includes the use of extensive personal protective equipment, which is believed to largely preclude exposure. Given the high air concentrations often associated with HazMat response, there is a great need for measurements of actual exposures inside masks or protective suits.

Given the paucity of actual exposure data, an assessment of the types of response can give us clues to firefighter exposure. An analysis of HazMat responses in Colorado, Iowa, Michigan, New Hampshire, New York, North Carolina, Oregon, Rhode Island, Washington, and Wisconsin from January 1, 1990, through January 1, 1992, was undertaken by the Agency for Toxic Substances and Disease Registry (ATSDR).[18] A total of 3,125 responses, thought to represent approximately 15% of all such incidents during that time period, were evaluated. The definition of hazardous substances was limited to the 200 substances on the ATSDR Superfund site list, all other insecticides and herbicides, and about a dozen other specific substances, mostly acids. Since petroleum products were specifically not included, the number of what would normally be defined as HazMat responses is underestimated by this study.

Analysis showed that 77% of the responses were to fixed locations and 23% were to transportation incidents. The mix of exposures faced by responders was much simpler than that of the fire environment. A single chemical was released in 88% of the incidents. More than five chemicals were released in only 6% of the incidents. In 466 responses in which someone was injured, more than five chemicals were released in only 2.4% of the incidents. Proportionally, fixed facility events had a greater fraction of the single chemical releases, although all 11 of the more than five-chemical releases also occurred in fixed facilities.

Volatile organic compounds (18%), herbicides (15%), acids (14%), and ammonia (11%) constituted the most commonly released substances. Thus, while the potential exists for exposure to a huge number of substances, in reality the universe of exposures is far smaller. A total of 2,501 people were injured during the releases studied; 200 of the injured were first-responders, which included facility personnel and firefighters. Of these first responders, 22% did not use any personal protective equipment, 43% used firefighter protective gear, 19% used level B respiratory protection, and 12% used level A respiratory protection.

Publications with quantitative measures of specific HazMat exposures are rare. In addition to typical firefighter concern about inhalation of combustion products, absorption of chemicals through the unprotected skin also may be a concern in HazMat incidents.[2]

REFERENCES

1. Anonymous: "Plastic fires" create new hazards for both firemen and public. JAMA 234:1211–1213, 1975.
2. Aufderheide TP, White SM, Brady WJ, Stueven HA: Inhalational and percutaneous methanol toxicity in two firefighters. Ann Emerg Med 22:1916–1918, 1993.
3. Atlas EL, Donnelly KC, Giam CS, McFarland AR: Chemical and biological characterization of emissions from a fireperson training facility. Am Ind Hyg Assoc J 46:532–540, 1985.

4. Barnard RJ, Weber JS: Carbon monoxide: A hazard to firefighters. Arch Environ Health 34:255–257, 1978.
5. Birky MM, Clarke FB: Inhalation of toxic products from fires. Toxic Organ Comp 57:997–1013, 1981.
6. Bollinger NJ, Schutz RH: NIOSH guide to industrial respiratory protection. Cincinnati, National Institute for Occupational Safety and Health, 1987, publication 87–116.
7. Brandt-Rauf PW, Fallon LF Jr, Tarantini T, et al: Health hazards of fire fighters: Exposure assessment. Br J Ind Med 45:606–612, 1988.
8. Brotherhood JR, Budd GM, Jeffery SE, et al: Fire fighters' exposure to carbon monoxide during Australian bushfires. Am Ind Hyg Assoc J 51:234–240, 1990.
9. Burgess WA, Sidor R, Lynch JJ, et al: Minimum protection factors for respiratory protective devices for firefighters. Am Ind Hygiene Assoc J 38:18–23, 1977.
10. Burgess WA, Treitman RD, Gold A: Air contaminants in structural firefighting. Boston, Harvard School of Public Health, 1979, NFPCA grant 7X008.
11. Davies RJ: Respiratory hypersensitivity to diisocyanates. Clin Immunol Allergy 4:103–123, 1984.
12. Demers PA, Checkoway H, Vaughan TL, et al: Cancer incidence among firefighters in Seattle and Tacoma, Washington (United States). Cancer Causes Control 5:129–135, 1994.
13. Dudley SF: Toxic nephritis following exposure to carbon tetrachloride and smoke fumes. J Ind Hyg 17:93–110, 1935.
14. Etzel RA, Ashley DL: Volatile organic compounds in the blood of persons in Kuwait during the oil fires. Int Arch Occup Environ Health 66:125–129, 1994.
15. Froines JR, Hinds WC, Duffy RM, et al: Exposure of firefighters to diesel emissions in fire stations. Am Ind Hyg Assoc J 48:202–207, 1987.
16. Gold A, Burgess WA, Clougherty EV: Exposure of firefighters to toxic air contaminants. Am Ind Hyg Assoc J 39:534–539, 1978.
17. Guidotti TL: Mortality of urban firefighters in Alberta, 1927–1987. Am J Ind Med 23:921–940, 1993.
18. Hall HI, Dhara VR, Price-Green PA, Kaye WE: Surveillance for emergency events involving hazardous substances — United States, 1990–1992. MMWR 43:1–6, 1994.
19. Halperin W, Landrigan PJ, Altman R, et al: Chemical fire at toxic waste disposal plant: Epidemiologic study of exposure to smoke and fumes. J Med Soc N J 78:591–594, 1981.
20. Hartzell GE, Packham SC, Switzer WG: Toxic products from fires. Am Ind Hyg Assoc J 44:248–255, 1983.
21. Henriks-Eckerman M-J, Engstrom B, Anas E: Thermal degradation products of steel protective paints. Am Ind Hyg Assoc J 51:241–244, 1990.
22. Heyer NJ, Weiss NS, Demers PA, Rosenstock L: Cohort mortality study of Seattle firefighters. Am J Ind Med 17:493–504, 1990.
23. Ives JM, Hughes EE, Taylor JK: Toxic atmospheres associated with real fires. Washington, DC, US Dept. of Commerce, National Bureau of Standards, 1972, NBS report 10 807.
24. Jankovic J, Jones W, Burkhart J, Noonan G: Environmental study of firefighters. Ann Occup Hyg 35:581–602, 1991.
25. Kaufman JD, Morgan MS, Marks ML, et al: A study of the cardiac effects of bromochlorodifluoromethane (Halon 1211) exposure during exercise. Am J Ind Med 21:223–233, 1992.
26. Large AA, Owens GR, Hoffman LA: The short-term effects of smoke exposure on the pulmonary function of firefighters. Chest 97:806–809, 1990.
27. Levine MS: Respirator use and protection from exposure to carbon monoxide. Am Ind Hyg Assoc J 40:832–834, 1979.
28. Levine MS, Radford EP: Occupational exposures to cyanide in Baltimore fire fighters. J Occup Med 20:53–56, 1978.
29. Lowry WT, Juarez L, Petty CS, Roberts B: Studies of toxic gas production during actual structural fires in the Dallas area. J Forensic Sci 30:59–72, 1985.
30. Materna BL, Jones JR, Sutton PM, et al: Occupational exposures in California wildland fire fighting. Am Ind Hyg Assoc J 53:69–76, 1992.
31. Materna BL, Koshland CP, Harrison RJ: Carbon monoxide exposure in wildland firefighting: A comparison of monitoring methods. Appl Occup Environ Hyg 8:479–487, 1993.
32. McMahon CK, Bush PB: Forest worker exposure to airborne herbicide residues in smoke from prescribed burns in the southern United States. Am Ind Hyg Assoc J 53:265–272, 1992.
33. Musk AW, Peters JM, Bernstein L, et al: Pulmonary function in firefighters: A six-year follow-up in the Boston Fire Department. Am J Ind Med 3:3–9, 1982.
34. Petajan JH, Voorhees KJ, Packham SC, et al: Extreme toxicity from combustion products of fire-retardant polyurethane foam. Science 187:742–747, 1975.
35. Radford EP, Levine MS: Occupational exposures to carbon monoxide in Baltimore firefighters. J Occup Med 18:628–632, 1976.

36. Reh CM, Letts D, Deitchman S: NIOSH health hazard evaluation report. U.S. Department of the Interior, National Park Service, Yosemite National Park, CA, 1994, HETA 90-0365-2415.
37. Rothman N, Correa-Villasenor A, Ford DP, et al: Contribution of occupation and diet to white blood cell polycyclic aromatic hydrocarbon-DNA adducts in wildland firefighters. Cancer Epidemiol Biomarkers Prev 2:341–347, 1993.
38. Sammons JH, Coleman RL: Firefighters' occupational exposure to carbon monoxide. J Occup Med 16:543–546, 1974.
39. Selala MI, Couke V, Daelemans F, et al: Fire fighting: How safe are firefighters. Bull Environ Contam Toxicol 5:325–332, 1993.
40. Schecter A: The Binghamton State Office Building PCB transformer incident: 1981–1987. Chemosphere 16:2155–2160, 1987.
41. Terrill JB, Montgomery RR, Reinhart RF: Toxic gases from fires. Science 200:1343–1347, 1978.
42. Treitman RD, Burgess WA, Gold A: Air contaminants encountered by fire fighters. Am Ind Hyg Assoc J 41:796–802, 1978.
43. Unger KM, Snow RM, Mestas JM, Miller WC: Smoke inhalation in firemen. Thorax 35:838–842, 1980.

VICTOR G. CIMINO, MD, DDS
SETH M. KROSNER, MD
MARELLA L. HANUMADASS, MD

BURN INJURIES IN FIREFIGHTERS

From the Division of Burn Services
  Department of Trauma
Cook County Hospital
Chicago, Illinois

Reprint requests to:
Marella L. Hanumadass, MD
Director, Sumner L. Koch Burn
  Center
Cook County Hospital
700 South Wood Street
Chicago, IL 60612

Despite improved safety in recent years, firefighting is considered one of the most hazardous professions. Between 1990 and 1993, more than 20,000 firefighters were burned in the line of duty, making such injuries a significant and obvious hazard for these individuals (Table 1). Firefighters face complex situations and unstable firegrounds that are fraught with peril. Flame and smoke themselves are chief among these potential dangers. In 1994 alone, half of all occupational deaths related to fire and explosion occurred amongst firefighters.[31] Fire and smoke (as opposed to falls or other injuries) are responsible for a large portion of death and injury and are in fact the single largest threat to life at the fireground. Sixty firefighters lost their lives at fires in 1994. Burns and smoke inhalation accounted for 22% of the fireground injuries incurred by firefighters in 1990–1993.[27,28] Clearly, the treatment of smoke inhalation and burn injury remains central to the occupational health of the firefighter.

EARLY MANAGEMENT OF MAJOR BURNS
Immediate Intervention

As with any traumatic injury, the treatment that burned patients receive in the initial period is critical. Immediate response in the field should include intravenous access and covering the patient to protect against hypothermia. The airway must be evaluated and controlled as necessary. If there is question of associated multiple trauma, the patient should be immobilized for transport to protect the cervical spine. The field treatment of a burn injury is, in fact, essentially the same as that of any other injured patient: emphasis is placed on rapid transport to an appropriate center.

TABLE 1. Firefighter Fireground Injuries, 1990–1993

Year	Total	Smoke Inhalation	Burns
1990	57,100	7,095	5,180
1991	55,830	7,525	4,960
1992	55,290	6,335	5,105
1993	52,885	5,540	5,990

From the National Fire Protection Association.

In the emergency center, the patient should be assumed to have suffered multiple injury until proven otherwise. The initial evaluation and treatment is the same for all victims of traumatic injury. The "ABCs" of trauma care focus on airway (A), breathing (B), and circulation (C). In other words, the need for intubation, ventilation, or fluid therapy should be addressed before attention is turned to the dramatic and distracting burn injury. It is not uncommon for physicians and nurses to inspect and dress the burn wound before it is clear that the patient has a stable airway and intravenous access. It is best to merely cover the patient with a clean sheet until more pressing issues have been addressed. A burn rarely kills a patient in the first hour or two after injury, an obstructed airway can kill in four minutes.

Evaluation of the Burned Patient

As soon as the immediate intervention is accomplished, a thorough and more in-depth evaluation must begin. A detailed history and physical examination should be performed. Attention must be paid to the circumstances of the burn (e.g., exposure to smoke) and details of treatment thus far. It is important to establish the presence of drug allergies and past medical problems early; it is impossible to predict if the patient will be intubated and unable to provide such information later. During the physical examination, assessing the extent and depth of the burn is of obvious importance. If extremities are burned, their neurologic and vascular status must be documented.

Estimation of the extent of a burn injury is simplified by the "rule of nines" (Fig. 1). This rule is easily remembered: each upper extremity and the head are 9% of the body's surface area, each lower extremity, the back, and anterior trunk are 18%.[23] The rule of nines is used to estimate the extent of burn injury in adults and is slightly inexact.

The depth of burn is assessed by observing the color and texture of the burn as well as the presence of blisters. An assessment of sensation in the wound is also helpful. Par-

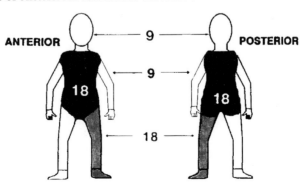

FIGURE 1. Rule of nines for estimating extent of burn injury.

TABLE 2. Differential Diagnosis of Depth of Burn

	Partial Thickness Burn	Full Thickness Burn
Sensation	Normal or increased sensitivity to pain and temperature	Anesthetic to pain and temperature
Blisters	Large, thick-walled; will usually increase in size	None or thin-walled; will not increase in size.
Color	Red, will blanch with pressure and refill	White, brown, black, or red; if red, will not blanch with pressure
Texture	Normal or firm	Firm or leathery

tially burned skin is painful, red, macerated, and exquisitely tender. Full-thickness burns, on the other hand, are leathery, not blistered, and asenate (Table 2). It is possible to have both partial- and full-thickness components in the same burn wound. Such wounds are called "mixed." It is also possible for deep-partial thickness burns to be converted into full-thickness injury by infection, inadequate resuscitation, or further local trauma. Full-thickness burns are created when the injury is of sufficient severity to cause coagulation necrosis of the skin, with permanent destruction of all living skin elements (Fig. 2). The depth of burn injury is determined by both the temperature and duration of the insult. Temperatures below 45° C (112°F) are well-tolerated indefinitely. Temperatures above 70° C (158° F) will destroy the skin entirely in less than 1 second.[30] Between these two temperature extremes, the degree of tissue destruction is a product of the temperature and duration of exposure. In other words, a temperature of 50° C will destroy skin more slowly than a temperature of 60° C.

Classification of Burn Severity

Burns are classified as major, moderate, or minor based on their extent, depth, etiology, and location. Associated injuries and other medical problems are also important (Table 3). A major burn has *any one* of the following characteristics: total extent of 25%

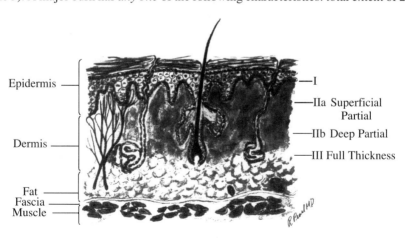

FIGURE 2. Cross-section of the skin shows skin appendages and depth of burn injury. (From Hanumadass ML, Kagan, R, Matsuda T: Management of pediatric burns. In Vidyasagar D, Sarnaik AP (eds): Neonatal and Pediatric Intensive Care. Littleton, MA, PSG Publishing, 1985, pp 215–226; with permission.)

TABLE 3. Severity Classification of Burns

Criteria	Major	Moderate	Minor
Total Burn	Over 25% (Adult)	15–25% (Adult)	Under 15% (Adult)
	Over 20% (Child)	10–20% (Child)	Under 10% (Child)
Full Thickness	Over 10%	2–10%	Under 2%
Location of Burns			
Face	Yes	No	No
Eyes	Yes	No	No
Hands	Yes	No	No
Feet	Yes	No	No
Perineum	Yes	No	No
Age	Over 60 years Under 2 years		
Associated Injury	Yes	No	No
Pre-existing Disease	Yes	No	No
Electrical Burn	Yes	No	No
Smoke Inhalation	Yes	No	No

From American Burn Association: Total care of burn patients, a guide to hospital resources. Bull Am Coll Surg 69(10):24–28, 1984.

or greater, total-full thickness extent of 10% or greater, associated smoke inhalation, or electrical cause. Any burns to vital or sensitive areas, such as the face, hands, feet or perineum, are also considered major burns. In contrast, moderate burns involve less than 25% but more than 15% of the body surface area, or have full-thickness components of 2–10% of body surface area. Any burn of lesser extent is considered minor. Any burns in poor risk patients (those older than 60 or younger than 2 years of age, and those with preexisting morbidities) are considered major regardless of their extent, depth, or location.[23] Any patient with a major burn should be cared for at a specialized burn center and should be transferred there as soon as possible.

DEFINITIVE BURN RESUSCITATION

The First 24 Hours

Adult patients with greater than 20% body surface area (BSA) burns require formal burn resuscitation. This extent of skin damage causes large volumes of extracellular fluid to be lost, resulting in profound hypovolemia. Fluid is lost from the body economy both in the burned tissue and throughout the body. In the burn site itself, the burn injury causes a great increase in microvascular permeability, resulting in the leakage of fluid and proteins.[19] There is also a generalized, systemic inflammatory response that results in cellular swelling and capillary leakage throughout the body.[10] Protein and fluid loss from the vascular space is greatest during the first 8 hours after injury.[4] "Burn shock" differs from hemorrhagic shock in that there is an ongoing loss of fluids for many hours after an injury, as opposed to a sudden loss of blood. In fact, a patient's vital signs are often normal at the time of the initial burn; severe hypovolemia may not develop for several hours. Fluid resuscitation is therefore designed to keep pace with ongoing losses. Hypoperfusion can convert marginal areas to full-thickness skin loss, making the resuscitation phase an important determinant of ultimate outcome.

The amount of resuscitation is dictated by the extent of burns and the body weight. The naked body weight should be measured immediately on admission. If significant treatment has occurred prior to this measurement, the weight will be inaccurate. In such cases it will be necessary to use the historical preburn weight (if the patient knows it). The fluid requirement of the first 24 hours is calculated suing the "Parkland formula."

$$4 \text{CC Ringer's lactate} \times \text{body weight in kg} \times \% \text{ BSA burned}$$

Half of the calculated requirements are given in the first 8 hours following injury; the remainder is divided over the ensuing 16 hours. The principal fluid used is lactated Ringer's solution (LR). Although slightly hypotonic, LR has a less deranging effect on the patient's pH when used in the enormous volumes often required by burn victims.[2] It is important to remember that the resuscitation is considered to have begun at the time of injury, and the goal is to administer half the required fluids in the initial 8 hours following the burn. For example, a man weighing 80 kg who has suffered a 45% burn will require $4 \times 80 \times 45 = 14,400$ cc of fluid in the first 24 hours. If his burn injury occurred only moments before, he will receive 7,200 cc over the next 8 hours, or 900 cc/hour. If, however, he was burned 4 hours ago and has received only 1,000 cc of fluids since that time, he will need 6,200 cc over the next 4 hours (1,550 cc/hour) to reach half of his requirements by 8 hours. If resuscitation is delayed at all, the patient may develop severe fluid deficits and require aggressive boluses of intravenous fluids. This formula is only a guideline; actual fluid needs may vary significantly. For example, patients who require fasciotomies generally increase their fluid needs by 20–30%, as do patients with electrical injury or smoke inhalation. Likewise, associated injuries (explosions, falls, car crashes) may cause hemorrhage, adding blood loss to the patient's fluid imbalance. Furthermore, the Parkland formula is only as accurate as the estimate of the body surface area involvement. For the 80-kg patient described, underestimating the burn by only 3% will result in a nearly 1 L underestimate of fluid needs ($80 \times 4 \times 3 = 960$).

The Second 24 Hours

Many burn centers add colloid to their resuscitation regimens in the second 24 hours of burn treatment. By this time, capillary leakage has largely resolved, and the serum protein levels will be reaching their lowest point.[4] It is much easier to maintain a hypoproteinemic patient's vascular volume with colloid solutions (like albumin, plasma, or dextran) than with LR or other crystalloids. The closure of the major capillary leaks by the second 24 hours makes the use of colloids more efficacious. Our institution infuses plasma or 5% albumin, 0.5 cc/kg/% burn, over the fourth 8-hour period (i.e., at the beginning of the second 24 hours). Furthermore, any required fluid boluses in the second 24 hours are given as colloid. The patient's maintenance intravenous fluid is changed to free water (D5W) during this period. At the conclusion of the second 24 hours after injury, parenteral or enteral nutrition is begun. A patient with a major burn injury will have greatly increased calorie and protein needs.

Monitoring the Resuscitation

Because the Parkland formula is designed to estimate fluid requirements, it is important to reassess the patient's response to treatment. This can be complicated.[24] Pain and fear will often artificially increase the blood pressure and the heart rate, making them unreliable predictors of volume status. This is particularly true in relatively young, healthy firefighters. These (presumably) physically fit individuals often can compensate for significant volume loss before they become hypotensive. Hypotension, in fact, can be a relatively *late* sign of trouble in these cases. More reliable is the patient's urine output. Patients who make 0.5 cc/kg/hour of urine, or more, are usually well hydrated. Patients with renal problems, diabetics, and those who take diuretics, however, may have misleading urine outputs. The patient's arterial blood pH should also be monitored. Hypoperfusion usually results in acidosis; an acidotic patient may well be under-resuscitated. Unfortunately, because there are other causes of acidosis, the pH may be inconclusive. In summary, there is no single variable that can be followed to assure an adequate response to treatment. The patient as a whole must be evaluated. An alert, normotensive patient with a normal arterial pH and 50–75 cc/hour of urine output is probably well-resuscitated. A confused, "shocky," acidotic, oliguric patient probably needs more aggressive fluid therapy. Many patients fall into the "gray zone" between these extremes. It is occasionally necessary to use invasive monitoring tools such as central venous pressure catheters or Swan-Ganz catheters to investigate volume status. It is less common to need such tools in burn injury than in other settings. This is because an acutely burned patient in shock is nearly always hypovolemic (rather than, for example, septic). In fact, a burn patient in shock should be assumed to need more fluid until proven otherwise. Furthermore, firefighters are typically healthy adults in good physical condition (indeed, otherwise healthy adults comprise the bulk of burn victims). It is difficult to cause fluid overload in such patients, particularly in the setting of major burns, making Swan-Ganz monitoring relatively superfluous in most cases. Nonetheless, the occasional patient with unclear response to aggressive treatment may benefit greatly from invasive monitoring.

Several complications may occur during the initial treatment phase. Some of the most common, like oliguria or hypotension, may be a symptom of inadequate resuscitation. Other problems include electrolyte imbalance (particularly hyponatremia) and hypothermia. Both of these are a result of the massive amounts of fluid that may be required.

SMOKE INHALATION

Firefighters are at dual risk for inhalation injury. They spend long times in close proximity to smoke. Although the advent of closed air packs reduces inhalation injury, smoke inhalation still accounts for a large percentage of injury and death. Smoke inhalation was the third leading cause of firefighter injury at the fireground scene in 1990–1993. Moreover, in 1994 alone, inhalation injury accounted for 31% of all firefighter deaths.[27]

Acute inhalation injury poses by far the greatest immediate threat to life. Early diagnosis and prompt treatment offer the best outcome. History is of paramount importance in the evaluation of inhalation injury (Fig. 3). Fortunately, firefighters can give reasonable and accurate accounts of inhalation exposure and severity. If the patient cannot give information regarding the fire, trained EMS personnel offer a reasonable alternative. The type of fire (house, industrial, chemical) and its location (indoor, outdoor) are important points. For example, some types of chemical fires pose unique risks because of fatal byproducts. Burned plastics and synthetics produce toxic and fatal gases: hydrochloric acid, phosphagene, chlorine, ammonia, cyanide, and others.[15] As much information as possible should be gathered regarding the fire to adequately assess the risk of inhalation injury.

Pathophysiology of Smoke Inhalation

A full discussion of the pathophysiology of inhalation injury is beyond the scope of this chapter. Instead, inhalation injury is broken down into three areas: (1) thermal injury, (2) anoxic injury, and (3) chemical injury. Thermal injury can be the most rapid and most devastating form of inhalation injury. Physical damage from heat injury is usually

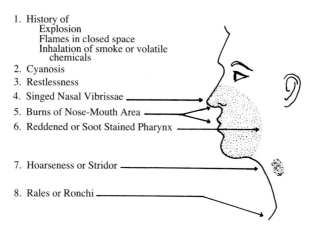

1. History of
   Explosion
   Flames in closed space
   Inhalation of smoke or volatile chemicals
2. Cyanosis
3. Restlessness
4. Singed Nasal Vibrissae
5. Burns of Nose-Mouth Area
6. Reddened or Soot Stained Pharynx
7. Hoarseness or Stridor
8. Rales or Ronchi

FIGURE 3. Clinical signs and symptoms of inhalation injury to the respiratory tract. (From Hanumadass ML, Kagan R, Matsuda T: Management of pediatric burns. In Vidyasagar D, Sarnaik AP (eds): Neonatal and Pediatric Intensive Care. Littleton, MA, PSG Publishing, 1985, pp 215–226; with permission.)

immediate. Direct heat can cause erythema, edema, and ulceration to the pharnyx and upper airways.[33] Thermal injury is usually limited to the anatomic region superior to, and including, the vocal cords. The physical arrangement of the upper airway is designed to allow cooling of the inspired air to protect the lower airways, making thermal burns of the lower airway rare. The efficacy of the upper airway as a heat as well as cold exchanger is well documented. An exception to this is superheated steam: steam can retain heat as it traverses the vocal cords to deliver a thermal burn to lower airways.[18] These thermal burns to the face, mouth, pharnyx, and vocal cords can be devastating if unnoticed. Rapidly developing pharyngeal and laryngeal edema during resuscitation can lead to loss of the airway. Once edema occurs, intubation can be difficult or impossible.

Physical examination of the face, airways, mouth, and oropharynx is a necessity in all suspected smoke inhalation victims. The presence of facial burns, singed nasal vibrissae, carbonaceous sputum, hoarseness, and stridor should give the examiner a strong suspicion of inhalation injury (see Fig. 3). All or any of these findings suggest inhalation injury and possible impending airway obstruction. If an upper airway burn is suspected, early prophylactic intubation, preferably nasal, should be performed. A wait-and-see approach can be risky and lead to an emergent and difficult intubation. This lethal scenario should be avoided at all costs.

Anoxic Injury

Anoxic injury can occur with inhalation victims. Very low levels of oxygen tension have been recorded at the scene of a fire. The oxygen concentration in some parts of a closed-space fire can plummet to as low as 2%.[11] These low levels of oxygen may lead to severe anoxic tissue damage to critical organs. This can prove fatal, even in early survivors. Most victims with extended anoxia perish at the scene of the fire. For this reason, oxygen must be administered to the burned patient as a first emergency measure. This most simple maneuver can greatly benefit inhalation victims.

Chemical Injury

Chemical injury to the airway is most often described as classic inhalation injury. The injury to the lung parenchyma and lower airways are not the result of direct heat, but, instead, a chemical insult. The incomplete products of combustion and carbon monoxide are the offending factors. The chemical insult can lead to edema and congestion of the mucosa of the tracheobronchial tree. This process can range from simple congestion and mild edema to desquamation and obstruction of the airways. The insult may be transient or may lead to full-blown adult respiratory distress syndrome (ARDS). A time line of inhalation injury exists, with risk for chemical injury peaking between 24 and 72 hours.

Carbon monoxide, a product of combustion, has a 200-fold greater affinity for hemoglobin than that of oxygen. This can lead to severe impairment of oxygen delivery and worsening hypoxemia.[34] (Table 4). Carboxyhemoglobin levels of greater than 20% cause confusion, headaches, and flushing, and greater than 60% can cause coma and death (Table 4).

Diagnosis

The diagnosis of smoke inhalation must be made largely by physical examination with a correlating history. Carboxyhemoglobin levels must be obtained and followed for a decreasing trend. If admission levels are greater than 20%, they must be followed hourly until levels fall below 20%.

Chest radiography may be of limited value early and yield no significant findings until several hours to days after exposure. Little to no radiographic chest abnormalities were noted in victims of the Las Vegas hotel fires at the time of hospital admission.[25] Reports such as this may question the usefulness of chest radiography at the outset. Nonetheless, chest radiography remains an important initial diagnostic test for two reasons. First, it can aid in the identification of any associated thoracic trauma. Second, an admission chest x-ray gives a useful baseline study. Any abnormalities on the admission film may alert the practitioner to an impending worsening course.

Arterial blood gases (ABGs) remain a fundamental study that must be performed in all suspected inhalation patients. Although the measurement of hemoglobin saturation may be inaccurate in the presence of carbon monoxide, the arterial blood gas is important in diagnosis. Arterial blood gases may differentiate between hypoxic (low O_2) and hypoventilatory (high CO_2) failure. Serial ABGs also are useful in following trends in respiratory failure. Bronchoscopy and xenon scans can be useful as diagnostic aids but should not be substituted for thorough physical examination. Physical examination and a high index of clinical suspicion remain the mainstay in the diagnosis of smoke inhalation.

Treatment

Treatment of smoke inhalation is largely expectant and supportive. If upper airway burns are suspected, immediate intubation should be performed. One hundred percent humidified oxygen must be administered immediately to all suspected inhalation vic-

TABLE 4. Carbon Monoxide Poisoning

Carbon monoxide-Hemoglobin	Signs and Symptoms
0–20%	No symptoms
20–40%	Headache, nausea, and vomiting
40–60%	Visual and intellectual impairment, decreasing consciousness
60–80%	Deep coma, respiratory depression, circulatory collapse
80+%	Death

tims. This not only helps correct hypoxemia but also reduces the half-life of caboxyhemoglobin. Administration of 100% oxygen for carboxyhemoglobminemia will decrease its half-life to 20–30 minutes from a normal half-life of 4 hours when breathing room air. If insult to lower airways or classic inhalation injury is present, intubation and mechanical ventilation also may be necessary. Positive end expiratory pressure (PEEP) may be necessary to prevent and treat oxygenation failure. This inhalation injury may evolve into ARDS, in which mechanical ventilation and high PEEP are needed for a protracted course of treatment. Bronchodilators may be useful in the presence of bronchospasm. Bronchospasm often accompanies the clinical inhalation picture and may appear early in its course. Prophylactic corticosteroids and antibiotics are not indicated for inhalation injury.[14] Corticosteroids may be useful in intractable bronchospasm but not in ARDS.[3,5,13,32] With protracted mechanical ventilation, well-directed antibiotics against specific organisms are useful in clinically proven pneumonias. Antibiotic choices should be based on culture and sensitivity data. There is no evidence to support prophylactic antibiotics in these patients.

Firefighters remain at high risk for smoke inhalation at the fireground. Although at risk for acute injury, whether chronic lung injury occurs is questionable.[22] Studies by Peters et al. have failed to confirm a previously reported chronic decrease in pulmonary function among firefighters.[26,29] This appears to be an area in need of further research.

CARE OF THE BURN WOUND

Initial Dressing

The initial care given to burn wounds is not definitive. No attention or energy should be diverted to the wound itself until immediately life-threatening issues are addressed. Until the patient's airway, breathing, and circulation are secure and all necessary intravenous access is obtained, the patient is simply covered in a dry, clean sheet to prevent gross contamination and heat loss. Referring institutions should not attempt to apply topical agents or definitive dressings. Doing so will delay transfer and cause further delays at the burn center when these dressings need to be removed.[23]

Definitive Dressing

Once the patient's initial evaluation and emergency treatments are completed and the patient is in a specialized burn facility, the definitive dressing can be placed. With extensive burns, this process is extremely labor intensive and usually requires the cooperation of a team of nurses and the physician. The wounds are washed and debrided. Dead skin is scrubbed or peeled away, and the wound is thoroughly cleaned of soot and debris. Hair-bearing areas are shaved. With partial-thickness burns, this process is agonizingly painful. Patients are sure to require aggressive doses of narcotic analgesia and must be closely monitored when the dressing is complete and the painful stimulation is ended. During this definitive dressing placement, careful attention is paid to the depth of injury and to the presence of peripheral pulses distal to circumferential burn wounds. Many centers use hydrotherapy to aid in this initial aggressive debridement. It is particularly important to remove dirt and particulate debris from a partial-thickness burn expected to heal without excision and skin grafting. If this is not done, the dirt will permanently discolor or tattoo the skin. After this thorough cleansing, silver sulfadiazine cream (or a similar topical agent) is applied generously to all affected areas, and the wound is dressed in sterile gauze and wrapped in rolled-type bandages. Burned extremities should be elevated and can be "hung" from intravenous poles. With circumferential burns of the extremities, the tips of the digits should be exposed to allow monitoring of the patient's circulation and sensation. A small hole should be left in the dressing to allow for palpation of the pulse. The dressings are changed on a daily basis.

Escharotomies and Fasciotomies

A full-thickness burn results in coagulation necrosis of the entire skin. The normally elastic skin is converted into a thick and inelastic eschar. If the wound is circumferential, or nearly so, this tough eschar can act as a tourniquet when the deeper tissues of the extremity swell. This can result in hydrostatic pressure within the extremity that exceeds capillary or even aterial perfusion pressures. This quickly results in a failure of circulation at the cellular level, followed by failure of blood flow to the entire distal limb. Such a situation is a surgical emergency. Peripheral nerves deprived of blood flow (and therefore oxygen) will suffer irreversible damage within 6 hours, muscle cells begin to die shortly thereafter. It is necessary to release the critically high pressures building within the extremity to prevent permanent destruction of the limb. This is done by means of an escharotomy, or incision into the eschar. Escharotomies are typically performed at the bedside with either cold blade or cautery. The eschar is slit longitudinally along the medial and lateral aspects of the arm or leg. The incision is made deep enough to fully release the tension of the eschar; it is usually necessary to incise down to bleeding tissue. Results are immediate: the soft, unburned fat and subcutaneous tissue force the eschar apart by as much as 2–3 cms, and the limb immediately becomes soft. There is obvious improvement in the color and capillary refill of the digits, and often the distal pulse will return.

If a burn is severe enough to cause edema of deeper tissues (electrical burns, in particular, are notorious for causing deep muscle damage), the inelastic fascia surrounding the muscle can cause increased pressure and decreased perfusion of the extremity. The mechanism is much the same as with the eschar. In such cases, fasciotomies must be performed to sharply divide the muscle fascia longitudinally. Care is taken when performing either escharotomy or fasciotomy to avoid the superficial peripheral nerves of the extremity. Upon completion of the escharotomy, blood rushes into the hypoxic areas. This results in the systemic infusion of toxins, because the products of ischemic metabolism are washed out of the affected muscles into the venous blood. These products include, but are not limited to, lactic acid and potassium. If there has been significant muscle damage, myoglobin also may be released into the circulation. For these reasons, brief periods of instability are occasionally seen shortly after escharotomy or fasciotomy. It should be emphasized that these procedures are extremely bloody; it is not unusual for several units of blood to be lost if long, multiple escharotomies are required. Escharotomies of the chest or trunk may be necessary to protect the patient's respiration when burns to the chest threaten to embarrass chest excursion. An early clue to this problem in intubated patients is a change in airway pressures and spontaneous tidal volumes. Decreases in these variables may be early clues to a stiffening chest wall. This may prompt chest escharotomies, typically in the axillary lines.

Surgical Management of the Burn Wound

In past years, many surgeons have traditionally disliked treating burns. Little wonder, since patients were often treated nonoperatively until they died. Their wounds either healed spontaneously or developed granulation tissue that would accept casually applied skin grafts. The idea that burns can be treated surgically at the outset is appealing to surgeons.[20] For the past 18 years, we have routinely practiced early surgical excision and wound closure. Based on our experience, we were able to define the purpose of early surgical excision and arrive at the following conclusions:[17]

1. Mortality from wound infections following major burns is virtually eliminated.
2. Mortality from other complications is lower when early excision and grafting decreases stress, hypermetabolism, and overall bacterial load, enabling patients to better resist other complications.
3. Small and moderate burns, less than 20% body surface area, can be safely excised and autografted in one sitting, with decreases in hospital stay, cost, and time lost from work. Early excision and grafting dramatically decreases the number of painful debridements and dressing changes required by all patients.
4. Scarring is less pronounced in wounds closed early, thereby leading to better appearance and the need for fewer reconstructive procedures.

Excision of an extensive burn wound is a major surgical undertaking and should be performed in a specialized burn center by an experienced surgical team. Support teams of anesthesiologists and nursing personnel are essential. Special equipment such as power dermatomes and skin meshers are needed to facilitate harvesting and expanding large sheets of skin. A well-stocked, cooperative blood bank is essential, as large volumes of packed red blood cells and other blood products are needed during these operations, which are associated with major blood loss. Intraoperative hemodynamic and respiratory monitoring and immediately available laboratory facilities are required during the procedure.

The aim of early excision is, of course, early wound closure. When the burn size is out of proportion to a limited donor area, permanent wound closure becomes problematic. Cultured epidermal cell layers, skin equivalent, and biodegradable temporary artificial skin are now being developed.[8] Even though some of these materials are now available for clinical use, their effectiveness is not fully proven. Further refinements in development and application are needed before routine clinical use of these materials. The best biologic dressing available for temporary wound closure is human cadaver homograft. Because of the difficulty in procuring cadaver skin, and also because of the fear of HIV transmission from human skin, various biosynthetic skin substitutes such as "Biobrane" have been developed for temporary closure of the burn wounds.

REHABILITATION

Advances in acute burn care and the development of specialized burn centers have reduced the mortality for any given total body surface area burn. Because patients with more serious burns now routinely survive, more challenging and intensive rehabilitation efforts have become necessary. Like acute burn care, rehabilitation care of these burn patients is best performed by an integrated team. For the most part, this rehabilitation team comprises the same members of the acute burn care team, but with modified roles. The individual members of the team must recognize their responsibilities in the total rehabilitation of burn patients. This is an ongoing process that begins when patients are admitted to the burn center. Burn patients require rehabilitation on several levels, for both body and spirit. Physical problems of decreased range of motion, hypertrophic scars, and contractures may be quite evident. Other disabilities such as emotional pain, self-image problems, and psychosocial issues may be less visible but no less disabling. It is for this reason that the rehabilitation phase of the burn patient encompasses both physical and psychosocial issues. Rehabilitation must be addressed comprehensively and concurrently. The goal of any rehabilitation program must be restore physical function and regain acceptance of a changed self-image.

Early Rehabilitation

As with any burn care plan, there are four goals to burn care and rehabilitation:[6]
1. The process should start early, preferably on the day of injury.

2. The care program should avoid prolonged immobilization. Any body part not moved actively should be moved passively.

3. Active motion should be started early, preferably on the day of injury.

4. There should be a planned program of activity and rehabilitation care each day. With consideration for the patient's level of consciousness, all joints should be actively moved each day. If the patient is intubated and sedated, passive motion must be performed on both the nonburned and burned areas. An organized, daily regimen with these specific goals is most favorable. Splinting must be instituted on patients when necessary. The maintenance of joints in favorable positions is of paramount importance. Intubated and sedated patients who are unable to cooperate with position and active motion will benefit from splinting. The goals of splinting are threefold:[7] to prevent deformity, to position joints away from naturally occurring contracture positions, and to improve and restore range of motion and function.

Scar and Contracture

The two most common types of postburn deformities seen are hypertrophic scar and burn contracture. Hypertrophic scar may occur on grafted and nongrafted areas. Pressure garments may aid in the shrinkage of hypertrophic scars as well as in reducing the amount of scar produced.[21] The scar may be excised with split-thickness skin grafting when it has fully matured. Contraction is a normal process of the healing wound. When this contraction of scar occurs across a joint space, a restriction in range of motion may occur. This often takes place in burned patients and can be addressed surgically when the contracture has matured, with either z-plasty or split-thickness skin grafting.

Psychosocial Issues

The psychosocial aspects of burns are complex. Approximately 25–30% of adults suffer some emotional problems 1 year after their burn.[1] Long-term emotional problems develop more frequently with burns to certain anatomic regions (such as the face and hand) as well as total body surface area burns greater than 30%.[9] The most common emotional problems encountered in burned patients are depression, anger, guilt, fear, and anxiety. These reactions must be recognized and dealt with on an individual basis. Anger may be misdirected by a patient toward staff or family. The root of this anger must be identified and addressed. The anger may have several targets, ranging from one's own self to coworkers. Resolution of this anger is a first step toward self-involvement in rehabilitation. Fear and anxiety are a part of any injury but are magnified with burns because of disfigurement and pain. Clear and open communication from nurses and physicians may aid in reducing anxiety.[16] Guilt may be present, especially in lone survivors of a disaster. This must be confronted early to facilitate patient involvement in recovery. Depression is common in burned patients and may be manifested in several ways—from overt crying to loss of appetite. Depression usually stems from feelings of helplessness. Patients must be urged to care for themselves as soon as possible. By walking and feeding themselves, patients feel less dependent. These tasks may keep patients focused and less apt toward depression. Because patients differ in preexisting psychiatric problems, family support systems, and coping mechanisms, the psychosocial issues of all patients must be individualized.

CONCLUSION

Firefighters face myriad hazards at the fireground scene. Principle among these dangers are the smoke and flame of the fire itself. Firefighters, like all burn victims, benefit from prompt and organized emergency treatment. Optimally, treatment should be rendered at a designated burn treatment center. A multidisciplinary burn team should be responsible for comprehensive care from admission to rehabilitation. This treatment goal optimizes survival and return to preburn activity.

Fortunately, the incidence of firefighter injury from smoke inhalation and burns is trending downward. These injuries are preventable in well-trained firefighters whose policies and procedures emphasize discipline and safety. Physical conditioning, training, periodic evaluation, and the use of specialized modern equipment can aid in the reduction of injury and death.

Acknowledgment

The authors gratefully acknowledge Ms. Nancy L. Schwartz, Fire Analysis and Research Division, National Fire Protection Association, Quincy, Massachusetts, for providing data on short notice. We also thank Ms. Lewandra Erwin and Ms. Dolores Serrano for typing and preparing this manuscript.

REFERENCES

1. Andreason NJC, Norris AS: Long term adjustment and adaptation in severely burned adults. J Nerv Ment Dis 154:352–362, 1972.
2. Baxter CR: Fluid volume and electrolyte changes in the early post-burn period. Clin Plast Surg 1:693–709, 1974.
3. Bernard GR, Harris T, Luce JE, et al: High dose corticosteroids in patients with the adult respiratory distress syndrome: A randomized double-blind trial. N Engl J Med 317:1565, 1987.
4. Birke G, Liljdahl SO, Plantin LO: Distribution and losses of plasma protein during the early stages of severe burns. Ann N Y Acad Sci 150:895–904, 1968.
5. Bone RL, Fischer C, Clemmer TP, et al: A controlled clinical trial of high dose methylprednisolone in treatment of severe sepsis and shock. N Engl J Med 317:653, 1987.
6. Boswick J: Comprehensive rehabilitation after burn injury. Surg Clin North Am 67:159–160, 1987.
7. Boswick J: Management of fresh burns of the hand and deformities resulting from burn injuries. Clin Plast Surg 4:621–631, 1974.
8. Burk JF, Quinby WC, Bondoc CC: Primary excision and prompt grafting as routine therapy for the treatment of thermal burns in children. Surg Clin North Am 56:477–494, 1976.
9. Chang FC, Herzog B: Burn morbidity: A follow-up study of physical and psychological disability. Ann Surg 183:34–37, 1976.
10. Civetta JM, Taylor RW, Kirby RR: Critical Care. Philadelphia, JB Lippincott, 1992.
11. Davies JWL: Toxic chemicals versus lung tissue—An aspect of inhalation injury revisited. J Burn Care Rehabil 7:213, 1986.
12. Demling RH: Burns. N Engl J Med 313:1389–1398, 1985.
13. Dressler D Skornik, Kupersmith S: Corticosteroid treatment of experimental smoke inhalation. Ann Surg 47–52, 1976.
14. Durtischi MB, Orgain C, Counts GW, et al: A prospective study of prophylactic penicillin in acutely burned hospital patients. J Trauma 22:11, 1982.
15. Dyer RF, Esch VH: Polyvinyl chloride toxicity in fires: Hydrogen chloride toxicity in firefighters. JAMA 235:393–397, 1976.
16. Hamburg DA, Adams JE: A perspective on coping behavior. Arch Gen Psychiatry 17:277–284, 1967.
17. Hanumadass ML: Recent trends in management of burns. Indian J Plast Surg 21:78–85, 1988.
18. Haponik E, Munster A: Respiratory Injury, Smoke Inhalation and Burns. New York, McGraw Hill, 1990, pp 48–49.
19. Harms B, Budai B, Smith M: Prostaglandin release and altered microvascular integrity after burn injury. J Surg Res 34:274–280, 1981.
20. Heimbach DM, Engrav LH: Surgical Management of the Burn Wound. New York, Raven Press, 1984.
21. Larson DL, Abston S, Evans EB, et al: Techniques for decreasing scar formation and contractures in the burned patient. J Trauma 11:807, 1971.
22. Loke J, Farmer W, Matthay R, et al: Acute and chronic effects of fire fighting on pulmonary function. Chest 77:369–373, 1980.
23. Martyn JAJ (ed): Acute Management of the Burned Patient. Philadelphia, WB Saunders, 1990.
24. Mason AD, Pruitt BA, Moncrief JA: Hemodynamic changes in the early post-burn period: The influence of fluid administration and vasodilator. J Trauma 1:36–46, 1971.
25. Miller EJ: Management of patients with smoke inhalation: The Las Vegas experience. In O'Donahue WJ (ed): Current Advances in Respiratory Care. Park Ridge, IL, American College of Chest Physicians, 1984.

26. Musk AW, Peters JM, Wagman DH: Lung function in fire fighters 1: A three-year follow-up of active subjects. Am J Public Health 67:626–629, 1977.
27. Karter M, LeBlanc P: U.S. firefighter injuries in 1993. Natl Fire Protection Assoc J 88(6):1994.
28. Washburn A, LeBlanc P, Fahy R: Fire fatalities in 1994. Natl Fire Protection Assoc J 89(4):1995.
29. Peters JM, Theriault GP, Fine LJ, et al: Chronic effects of fire fighting on pulmonary function. N Engl J Med 291:1320–1322, 1974.
30. US Consumer Product Safety Commission: Special report: Accidents and injury involving scald burns from tap water sources. Bethesda, MD, US Consumer Product Safety Commission, 1978.
31. US Department of Labor, Bureau of Labor Statistics: Fatal work place injuries in 1993: Collection of data and analysis Washington, DC, Bureau of Labor Statistics, 1995, report 891.
32. The Veterans Administration Systemic Sepsis Cooperative Study Group: Effects of high dose glucocorticoid therapy on mortality in patients with clinical signs of systemic sepsis. N Engl J Med 317:659, 1987.
33. Wald PH, Balmes JR: Respiratory effects of short-term, high intensity toxic inhalation. Smoke, gases and fumes. J Inten Care Med 2:260, 1987.
34. Zikria BA, Weston GC, Chodoff M, Ferrer JM: Smoke and carbon monoxide poisoning in fire victims. J Trauma 12:641–645, 1972.

KENNETH E. BIZOVI, MD
JERROLD D. LEIKIN, MD

SMOKE INHALATION AMONG FIREFIGHTERS

From the Division of Occupational
  Medicine
Cook County Hospital (KEB)
  and
Rush Presbyterian-St. Luke's
  Hospital (JL)
Chicago, Illinois

Reprint requests to:
Kenneth E. Bizovi, MD
Division of Occupational Medicine
Cook Country Hospital
720 South Wolcott
Chicago, IL 60612

Smoke inhalation may account for up to 75% of fire-related deaths.[16,38] Inhalation injury can present with a wide variety of complaints and findings. Patients may have multiple associated traumatic injuries and burns. While about 10% of patients with less than 40% total body surface burns will have inhalation injury, more than half the patients with greater than 80% burn will have inhalation injury.[10] Inhalation injuries in burn patients carry a mortality rate of 45–78%.[18] A thorough review of smoke inhalation is published elsewhere in this series.[31] The pathophysiology of this broad category of injury is still being researched. Many controversies in the management of the smoke inhalation patient still exist. Consideration of many toxicologic, physiologic, and pulmonary processes is necessary to manage victims of smoke inhalation.

Smoke contains the gaseous, liquid, and solid products of thermal degradation. The content of smoke depends on the material, the temperature, additives, and the presence of oxygen.[31] Although the clinician will rarely know the exact substances in a fire, knowledge of the content of smoke will help illustrate the patterns of injury from smoke inhalation. The components of smoke can be broken down into simple asphyxiants, chemical asphyxiants, and irritants. Simple asphyxiants cause hypoxemia by displacing oxygen and decreasing the FiO_2. Examples of simple asphyxiants include carbon dioxide, methane, helium, nitrogen, and nitrous oxide.[3] Chemical asphyxiant such as carbon monoxide, cyanide, hydrogen sulfide, and arsine gas lead to hypoxia by interrupting the transport or use of oxygen. Irritants can lead to bronchospasm, inflammation, and destruction of pulmonary tissue. Common irritants found in smoke include acrolein, ammonia, sulfur dioxide, hydrogen chloride, chlorine gas, and phosgene.[22]

PATHOPHYSIOLOGY

There are three primary mechanisms of injury in smoke inhalation: asphyxiation, thermal damage, and pulmonary irritation. Asphyxia can be from primary lack of oxygen in fire, carbon monoxide poisoning, cyanide poisoning, or more rarely, production of methemoglobinemia. Thermal inhalation injury is from hot gases produced by fire and primarily affects the upper airway. The pathophysiology of pulmonary irritant injury includes direct caustic injury, which is usually from volatile products of fire and inflammatory reaction to inhaled gasses and particulate matter.

The physiologic consequences of hypoxia, heat injury, and pulmonary irritants define the progression of systemic and pulmonary injury in the smoke inhalation victim. Coma and cardiovascular compromise can be the result of significant anoxia. Upper airway edema from heat injury can lead to airway compromise. Inhalation of irritant gases and particulate matter can cause a wide range of pathologic processes and clinical sequelae that account for many of the acute and delayed effects of smoke inhalation.

Heat and Upper Airway Injury

Inhalation of hot gases can lead to direct thermal injury of the upper airway. Airway compromise typically occurs from edema of injured mucosal surfaces. The upper airway dissipates heat effectively and protects the subglottic airway. Most thermal injury is limited to the supraglottic airway. Exceptions to limited thermal injury occur with the inhalation of steam, explosive gases, and aspiration of hot liquids.[17]

Irritants

Irritants can cause direct tissue injury, acute brochospasm, and inflammatory response. The direct injury is frequently the consequence of the chemical's acid base status. Ammonia produces alkaline injury. Sulfur dioxide and chlorine gas lead to acid injury. Other chemicals have other mechanisms of direct injury; for instance, acrolein causes formation of free radicals and protein denaturation.[22] The location of injury depends on the solubility of the substance in water. High-solubility substances such as acrolein, sulfur dioxide, ammonia, and hydrogen chloride cause injury to the upper airway. Substances with intermediate solubility such as chlorine and isocyanates cause upper and lower respiratory tract injury. Phosgene and oxides of nitrogen have low water solubility and cause parenchymal injury.[22,37]

Airway obstruction and alveolar injury can occur from a number of mechanisms. Direct injury to the tracheobronchial tree can lead to mucosal edema, sloughing of epithelium, bronchorrhea, and exudation of proteinaceous fluid.[37] This can be exacerbated by inflammatory mediators leading to increased blood flow and release of proteolytic enzymes from polymorphonuclear leukocytes. Bronchospasm can lead to further obstruction of air flow. Secretions combined with the solid components of smoke can cause mechanical obstruction of airways. Interstitial edema and fluid accumulation of fluid in the alveoli can cause adult respiratory distress syndrome. Immediate alveolar collapse leading to increased alveolar arterial oxygen gradient may be mediated by denaturation of surfactant.[30]

Asphyxiants

Tissue hypoxia can occur secondary to several mechanisms. Decreases in inspired oxygen can occur due to consumption of oxygen through combustion and displacement of oxygen by the products of combustion and pyrolysis. Decreased oxygen-carrying capacity of the blood most frequently results from carbon monoxide, but can also be from methemoglobinemia. Poor oxygen delivery can occur form circulatory compromise, which can be from hypovolemic shock or cardiotoxic sequelae of smoke inhalation. Cyanide, a common product of combustion, causes cellular hypoxia by poisoning aerobic metabolism.

The environment in a fire rapidly becomes devoid of oxygen. The production of CO_2 consumes and displaces O_2 during combustion, especially in closed spaces. In models of fire simulation in small spaces with high volume of flammable material, consumption of oxygen and production of CO_2 pushed FiO_2 to 3%.[29] When a room bursts into flames, the FiO_2 may drop from 21% to 15%.[13] The decrease in FiO_2 leads to hypoxia despite adequate circulation and oxygen-carrying capacity. Because of their relatively high metabolic rate and high oxygen extraction, the cardiovascular and central nervous system are most susceptible to hypoxic stress. This can lead to mental status changes at the time of exposure and is one of many reasons that people are not able to escape smoke-exposed areas.

Carbon monoxide causes asphyxiation by decreasing the oxygen-carrying capacity of the blood. It binds to hemoglobin with more than 200 times the affinity of oxygen. Carbon monoxide causes a shift to the left of the oxyhemoglobin dissociation curve, preventing the release of oxygen at the tissue. This occurs because when one carbon monoxide is bound to a hemoglobin tetramer, the binding of oxygen to the three remaining heme molecules becomes tighter. Because of the left shift, a carboxyhemoglobin level of 10% will decrease oxygen delivery more than a 10% decrease in hemoglobin from anemia. Carbon monoxide has been shown to bind to the cytochrome chain in vitro.[33] Theoretically this would interfere with aerobic respiration at the cellular level. The clinical significance of this effect is uncertain.

The clinical signs and symptoms in carbon monoxide poisoning are nonspecific. As with other asphyxiants, the primary targets are the nervous and cardiovascular systems. Concentrations of carbon monoxide more than 100 ppm can cause death within minutes. At low levels, patients may complain of nausea, headache, and dizziness. There may be no findings on physical or routine neurologic examination. Neuropsychiatric testing will more frequently demonstrate memory loss and more subtle decrease in cerebral function.[42] As neurologic compromise increases, headache may worsen and complaints of difficulty concentrating and weakness may occur. Confusion, ataxia, cognitive deficits, and focal deficits may be noted. Patients can develop coma and seizures in more severe cases. Cardiovascular effects are the consequence of the increased work required to make up for decreased oxygen delivery and direct cardiotoxicity. The cardiotoxic effects are only partially explained by the decrease in oxygen delivery to the heart. The binding of carbon monoxide to myoglobin and poisoning of mitochondrial cytochrome oxidase are other proposed mechanisms of direct cardiotoxicity. Cardiac effects may be manifested by palpitations, chest pain, shortness of breath, dyspnea on exertion, tachypnea, and tachycardia. Symptoms of ischemia and electrocardiogram changes may occur. Severe cardiac toxicity is manifested by hypotension. This may play a significant roll in the development and exacerbation of neurologic sequelae.

The area of greatest concern in the literature is the ability of carbon monoxide to cause delayed neurologic sequelae and the ability of various therapeutic modalities to prevent poor outcomes.[36,41,42,47,49] Neurologic consequences of carbon monoxide poisoning include personality and memory disturbances, a Parkinson-like disorder, mixed motor sensory peripheral neuropathy, and psychiatric disturbances.[38] The current absence of a controlled randomized trial of hyberbaric treatment versus 100% normobaric oxygen leaves the subject open to debate.

Combustion of plastics, polyurethane, wool, silk, nylon, nitrocellulose, polyacrylnitriles, rubber, and paper products can lead to the production of cyanide gas.[22] Cyanide is a chemical asphyxiant that interferes with cellular metabolism by binding to the cytochrome oxidase system and halting aerobic metabolism. The actual binding site is the ferric (Fe^{3+}) ion on cytochrome A_3. The manifestations of poisoned aerobic metabolism are lactic acidosis from anaerobic metabolism and decreased oxygen consumption. Although the mechanism of cyanide poisoning is different from that of carbon monoxide, the clinical manifestations are quite similar. The central nervous system and cardiovascular system are affected most often, but the signs and symptoms are nonspecific. Central nervous system findings include agitation, cephalgia, disorientation, seizures, coma, and death. The cardiovascular findings include tachycardia, bradycardia, and hypertension that progresses to hypotension with bradycardia and cardiovascular collapse. There is no method for rapid detection of cyanide. The nonspecific clinical findings and the inability to establish definitive diagnosis rapidly is further complicated by the rapid onset and progression of toxicity requiring immediate treatment. Cyanide should be considered in all smoke inhalation victims with central nervous system or cardiovascular findings. The most practical approach would be to assess for other causes of clinical findings such as carbon monoxide, hypoxia, and hypoglycemia. If no other explanation can be established, treatment for cyanide should be considered. In smoke inhalation victims with carbon monoxide poisoning, failure to improve with administration of oxygen may indicate coexistent cyanide and carbon monoxide poisoning.

Methemoglobinemia occurs in fire due to heat denaturation of hemoglobin, oxides of nitrogen produced in fire, and methemoglobin forming materials such as nitrite. Methemoglobinemia is less frequently reported than carbon monoxide or cyanide poisoning.[16] The formation of methemoglobin involves the conversion of the iron from the ferrous (Fe^{2+}) to the ferric (Fe^{3+}) state. The pathophysiologic consequence of methemoglobin formation is a decrease in the oxygen-carrying capacity of the blood and a shift of the oxyhemoglobin dissociation curve to the left similar to carboxyhemoglobin. The clinical effects of methemoglobin depend on the color of methemoglobin and the development of hypoxia. Methemoglobin causes more cyanosis than deoxyhemoglobin. A concentration of 1.5 gm/dL of methemoglobin will cause cyanosis, versus 5 gm/dL of deoxyhemoglobin.[34] Patients with methemoglobinemia may have cyanosis out of proportion to respiratory distress. The consequences of methemoglobinemia depend on the concentration of methemoglobin, the patient's underlying cardiovascular and respiratory status, and anemia. Symptoms of moderate toxicity include dyspnea, cephalgia, dizziness, syncope, and constitutional complaints. More severe symptoms include dysrhythmias, seizures, and central nervous system depression.[34]

EVALUATION AND MANAGEMENT

Smoke inhalation can present with a broad spectrum of severity and with multiple associated problems. Inhalation injury can range from immediate threat to airway and respiration to minor irritation. Smoke inhalation is often associated with burns but does not have to be. Patients in fires are subject to traumatic injury by a variety of mechanisms, including; jumping, falls, and falling objects. Evaluation and treatment should start with prehospital care personnel. Initial evaluation should proceed rapidly with the goal of stabilizing critically ill patients. The primary survey coupled with a history directed at identifying risk factors for smoke inhalation can further sort out patients with noncritical cases into high and low risk for smoke inhalation. A thorough examination after the patient is stabilized can identify all injuries and be used to guide further therapeutic and management decisions.

Primary Survey

Simultaneous assessment and treatment of critical issues should be undertaken as in trauma victims.[35] Intubation for airway control should be done in patients with no gag, with clinical signs of respiratory compromise, stridor, or significant traumatic chest injury. Cervical immobilization should be maintained in any patient with abnormal mental status or signs of head trauma. Breathing should be assessed by respiratory rate, chest wall motion, and auscultation for air movement. Circulation can be rapidly assessed by blood pressure, pulse rate, level of consciousness, skin color, capillary refill, and palpation of pulses. Intravenous access with two large-bore catheters should be obtained in all critically ill patients. Brief neurologic evaluation can document disability by noting whether the patient is alert, responsive to verbal stimuli, responsive to painful stimuli, or unresponsive. Patients with an altered level of consciousness should have empiric administration of glucose or rapid measurement of glucose, oxygen by a non-rebreather mask, naloxone 0.4–2.0 mg IV push, and thiamine 100 mg IV push. All clothes must be removed to expose traumatic injuries, burns, and burning clothing that may be causing prolonged thermal injury. This part of the assessment should include examination of the patient's back. In patients requiring cervical spine immobilization, this can be accomplished by logrolling. Long-bone fractures should be splinted immediately.

History

Historical information that points to high risk of smoke inhalation should be sought. Closed-space fires increase the risk of significant smoke inhalation. Particular materials in fire may point to certain asphyxiants. Polyurethane, wool, or silk increase the risk of cyanide toxicity. Patients in enclosed areas have an increased incidence of respiratory injury.[32] The patient's condition at the scene may yield critical information, such as the loss of consciousness or death of people in the same environment. Past medical history of respiratory disease such as asthma or COPD will predispose patients to respiratory findings.

Secondary Survey

Any patient with signs or symptoms of respiratory compromise should be considered to have inhalation injury and undergo evaluation appropriate for the specific findings. Paramount in these patients is identification of impending respiratory failure. Hoarseness, change in voice, complaints of throat pain, and painful swallowing indicate an upper airway injury that may be severe. One author suggests that carbonaceous sputum "should be regarded as a marker of exposure."[16] Other signs of respiratory injury include tachypnea, wheezing, rales, rhonchi, and use of accessory muscles. Patients with facial burns should be carefully evaluated for smoke inhalation. Burns involving the nose, brows, lips, and neck are at high risk for inhalation injury. Phillips et al. found a 59% incidence of respiratory injury compared with 22% incidence in patients with peripheral facial burns or no facial burns.[32] The incidence of pneumonia has been shown to correlate with the presence of facial burns.[50] Those with facial burns showed an increased mortality and a trend toward increased need for ventilatory support.[50] Large cutaneous burns indicate an inability to escape flame and should be considered risk factors for respiratory sequelae.[16]

Signs of head, chest or abdominal trauma should be looked for and monitored. About 60 of 114 fatalities in wildland firefighters from 1981 to 1990 were due to either heart attack or internal trauma.[9] Almost 75% of these deaths occurred during fire suppression activities.

Fluids

Optimal fluid administration is not established. The clinician must attempt to maintain adequate circulation and perfusion without unnecessarily increasing pulmonary hydrostatic pressure. Monitoring of vital signs, urine output, and acid base status can be used to evaluate perfusion. Following chest radiograph, an A-a gradient can be used to assess the pulmonary status. If the patient develops adult respiratory distress syndrome (ARDS), or pulmonary edema, a Swan-Ganz catheter can be used to assist fluid management.

LABORATORY AND DIAGNOSTIC STUDIES

Diagnostic testing can be used to confirm or exclude clinically suspected injury, document exposure, and establish baseline values, which can be used to follow trends. Patients with a history of exposure to smoke should have laboratory tests to document carboxyhemoglobin and oxygenation. Patients with respiratory complaints or findings and patients requiring intubation should have a chest x-ray. Other ancillary tests may assist in the evaluation of the exposed patient and should be obtained as indicated.

Evaluating Oxygenation

Pulse oximetry and arterial blood gases (ABGs) can be used to evaluate oxygenation. The pulse oximeter is rapid, noninvasive, often readily available in the emergency department, and may be the most rapid assessment of oxygenation. These machine can be misleading in the setting of carbon monoxide exposure or methemoglobinemia.[4,37,45,48] Carboxyhemoglobin is measured as oxygenated hemoglobin giving a falsely elevated saturation. Methemoglobinemia may show depressed saturation, but the decrease does not accurately reflect the level of methemoglobin. As levels of methemoglobin reach 30% and higher, the pulse oximeter will not go below 85%. Knowledge of these limitations will allow clinicians to use the information from these machines appropriately. The presence of a low saturation in any setting requires some investigation to determine the cause. Arterial blood gases can be used in these situations to give a correct measured saturation, carboxyhemoglobin, and methemoglobin levels. The arterial oxygenation can be used to calculate the alveolar arterial oxygenation gradient (A-a gradient). An elevated A-a gradient may indicate shunting from either acute injury or chronic disease. Shunting may be present from plugging of bronchi, atelectasis, alveolar injury, ARDS, or pulmonary edema. One other early indicator of pulmonary injury is the ratio of arterial PO_2 to the percent of inspired oxygen (F), or the P:F ratio. The P:F ratio normally is over 400. A ratio of less than 300 suggests pulmonary injury, and a ratio of less than 250 may be an indication for vigorous pulmonary therapy.[18] Blood gases give further insight into the patient's condition by giving information about acid base status and ventilation. A normal blood gas does not rule out smoke inhalation. The delayed inflammatory sequelae may lead to late findings of hypoxia and an increase in the arterial alveolar gradient. Arterial blood gases may be used to identify the presence and progression of inflammatory response later in the course of exposure.

Chemistry

The chemistry laboratory can be used to obtain electrolytes, BUN, creatinine, and other tests for specific indications. Electrolytes can identify anion gap and indicate acid base status. Measurement of lactate may be a source of metabolic acidosis secondary to carbon monoxide, cyanide, methemoglobinemia, or hypoxia. Lactate levels of greater than 10 mmol/L have been shown to be a sensitive indicator of cyanide levels greater than 1 µg/ml in victims of smoke inhalation.[2] BUN and creatinine can used to screen renal function and establish a baseline in patients in shock or with rhabdomyolysis. Patients with large cutaneous burns, crush injuries, or obtundation for long periods should have serum CPK and urine myoglobin measurements. Urinalysis may give the earliest evidence of rhabdomyolysis if blood is reported and there are no red blood cells on the microscopic analysis. This occurs because urine myoglobin is detected as blood on most urinalysis tests. If rhabdomyolysis is suspected, the urine should be sent for myoglobin testing.

Radiologic Studies

A chest radiograph should be obtained in patients with respiratory complaints or findings as well as any intubated patient. Most chest radiographs are normal after smoke inhalation with a false negative rate as high as 92%.[37] They can show atelectasis, pulmonary edema, and ARDS. Chest radiographs also can be used later to follow the progression of lung injury, identify pneumonia, and show pulmonary edema or ARDS.

Depending on the setting of patient management, radionuclide ventilation scans can be used to evaluate patients for evidence of smoke inhalation. It is used in some referral centers but not in many community settings.[16] "The inhomogenous uptake of radionuclide, and its focal retention and delayed clearance (e.g., > 90 s), characterizes areas of distal airway obstruction and air-trapping due to tracheobronchial injury."[16] Ventilation-perfusion scans are still used primarily to evaluate patients for suspected pulmonary embolism. The radionuclide, 99m Tc-diethylenetriamne pentacetate (DPTA) has been used experimentally to assess alveolar injury by measuring alveolar epithelial permeability. The leak of this tracer has been demonstrated in animals and some humans with smoke inhalation injury.[8] The clinical usefulness of this study has not been established.

Computed tomography of the chest is usually not indicated in the acute evaluation of smoke inhalation victims but can be useful later. The CT can be used to evaluate patients for tracheal stenosis, empyema, or chronic sequelae such as bronchiolitis obliterans or bronchiolitis.[16] It is more sensitive than xenon-133 scanning or chest radiography in detecting atelectasis.[37]

Assessment of the Upper Airway

Clinicians managing victims of smoke inhalation need to maintain frequent monitoring of the upper airway. Intubation should be considered in patients with stridor, use of accessory muscles, or retractions. Early intubation should be performed if (1) smoke inhalation is accompanied by a decreased level of consciousness, (2) posterior pharyngeal edema, (3) full thickness nasolabial burns or (4) circumferential neck burns.[10] Endoscopic evaluation of the airway can be considered in patients who are at high risk for inhalation injury and who have not already required intubation. A high (96%) correlation has been noted between positive bronchoscopic findings and the clinical triad of closed space fire, carboxyhemoglobin over 10%, and carbonaceous sputum. If any of these findings are present, bronchoscopy should be performed.[37] Bronchoscopy can be accomplished rapidly, is well tolerated by patients, can have both diagnostic and therapeutic value, and provides immediate information regarding the degree of injury to the airway. Direct visualization may identify injuries that require emergent intubation. Lower airway injury is less likely in the presence of a normal upper endoscopy, but it does not rule out distal injury. Intermediate and low solubility irritants such as chlorine, isocyanates, oxides of nitrogen, and phosgene can cause lower airway injury greater than upper airway injury. Fiberoptic bronchoscopy can be used to assist intubation by acting as a guidewire for an oral tracheal tube. Endotracheal intubation may need to be maintained until upper airway edema decreases (usually within 4 days after injury). Serial evaluations for edema and extent of healing can be used to assess the need for continued intubation.

Assessment of the Lower Airway

Endoscopic evaluation of the subglottic airway must be weighed against other diagnostic modalities such as chest x-ray, pulmonary function tests, ventilation-perfusion scans, and arterial blood gases. The presence of inhalation injury to the trachea and main bronchi can be definitively diagnosed by this procedure. This procedure can be used to remove mucous and debris plugs. Bronchial alveolar lavage has been used to monitor chemical and cellular events of smoke inhalation but the information is not clinically useful.[6] Bronchoscopy can be used to diagnose tracheal or bronchial stenosis and to acquire tansbronchial biopsy. Biopsies can be used to diagnose inflammatory process such as bronchiolitis obliterans and diffuse lung disease unrelated to smoke inhalation.

TREATMENT

Bronchodilators

The administration of bronchodilators and antiinflammatory agents has been considered in the management of smoke inhalation. The effectiveness of bronchodilators is not well documented. Nebulized beta-agonists can be used and discontinued if no improvement is noted.[7] Patients with underlying COPD or asthma may obtain more benefit from beta-2 agonists.[16] Humidified oxygen or air has been used to prevent secretions from drying.

Management of Carbon Monoxide Poisoning

Treatment for carbon monoxide poisoning involves supportive measures and administration of normobaric or hyperbaric oxygen. Oxygen should be administered by non-rebreather face mask during prehospital care or in the emergency department as soon as smoke inhalation is recognized or suspected. Evaluation for suspected carbon monoxide can be accomplished by venous or arterial measurement with a co-oximeter of a blood gas machine.[45] Pulse oximetry will show high saturations despite the presence of high carboxyhemoglobin levels and is not adequate to assess the presence of carbon monoxide.[4,48] The halflife of carbon monoxide on room air at 1 atmosphere is 4–5 hours. With the administration of 100% normobaric oxygen the halflife decreases to 40–80 minutes; with 100% normobaric oxygen at 2.8 atmospheres the halflife is further reduced to 20–50 minutes.[38,43]

Pregnant patients require special consideration in the setting of carbon monoxide poisoning. Fetal hemoglobin has a greater affinity for carbon monoxide, leading to fetal levels above maternal levels and prolonged elimination time.[23] Pregnant patients should receive treatment with normobaric oxygen for five times longer than it takes to eliminate the carbon monoxide from the maternal circulation.[26] Hyperbaric oxygen should be considered for levels greater than 15% in pregnant women, but there is variation in this criterion among centers.[43,46]

The biggest controversy surrounding the management of carbon monoxide poisoning involves the use of hyperbaric oxygen.[41] Hyperbaric oxygen at 2 or more atmospheres will increase oxygen delivery and decrease carbon monoxide elimination to a greater extent than normobaric oxygen, giving hyperbaric oxygen the theoretic advantage. A definitive study of hyperbaric oxygen versus normobaric oxygen has not been done. The efficacy of hyperbaric oxygen in preventing delayed neurologic sequelae is not established.[36,42,47] There are multiple sources of criteria for hyperbaric oxygen in the setting of carbon monoxide poisoning.[11,43,44,46] Clinicians should be familiar with the available hyperbaric facilities and contact them to make a cooperative decision about treatment and the need for hyperbaric therapy. The Divers Alert Network (DAN) at Duke University can assist in locating hyperbaric facilities. When transfer is required, the risks of transfer must be weighed against the benefits of hyperbaric oxygen. Indications for hyperbaric oxygen include altered mental status, neurologic findings, cardiovascular dysfunction, pulmonary edema, severe acidosis, and loss of consciousness.[46] Other criteria for considering HBO include carboxyhemoglobin greater than 25%, history of cardiovascular disease and age older than 60, and pregnancy with symptoms or carboxyhemoglobin level greater than or equal to 15%.[23,46]

Management of Suspected Cyanide Poisoning

The most difficult aspect of treating cyanide poisoning in the setting of smoke inhalation is deciding when to initiate treatment. The clinical signs of cyanide intoxication are nonspecific and include anxiety, flushing, dizziness, and headache.[51] There is no rapid test for the presence of cyanide. No clinical or laboratory test has been described that correlates specifically with significant cyanide exposure. There are theoretical concerns about treating a patient with carbon monoxide poisoning for presumed cyanide intoxication because therapy involves the formation of methemoglobin, which will further inhibit oxygen carrying capacity. Conversely, cyanide exposure during fire has been documented.[2,24,39] Cyanide levels have been found to be elevated in 78–100% of fire fatalities.[1,39,40] In one study, 90% of the fire survivors had cyanide levels greater than 0.1 µg/ml.[39] Studies that looked at the correlation between cyanide and carboxyhemoglobin levels have weak or no correlation.[24,39] Some studies have found a significant correlation between carboxyhemoglobin levels and cyanide levels.[2,5,20]

In the absence of rapidly available cyanide levels, the ability to treat empirically should be maximized. An ideal antidote would be highly effective in rapidly treating cyanide with no adverse effects in the setting of carbon monoxide poisoning or smoke inhalation injury. If such an antidote were available, one could argue for empiric treatment of any smoke inhalation victim who is critically ill or perhaps even moderately ill. In the setting of a highly effective antidote with potential side effects in smoke inhalation victims, the priority would be on defining criteria for which the benefits outweigh the risks.

The management of cyanide involves creating an alternative site for the binding of cyanide to compete with the cytochrome oxidase binding site and providing substrate necessary to convert cyanide to a more nontoxic metabolite. The most frequently used therapy is a cyanide antidote package that contains three antidotes—amyl nitrite, sodium nitrite, and sodium thiosulfate—in quantity sufficient to treat two people. Both the amyl nitrite and the sodium nitrite induce a methemoglobinemia that binds cyanide. The major disadvantage of the nitrites is hypotension secondary to vasodilatation. The original goal of nitrite therapy was to create a methemoglobin level of 20–30%. In a series of seven patients who were given the standard 300 mg dose of sodium nitrite, the methemoglobin levels ranged from 7.3–13.4%; however, these levels do not include measurement of cyanomethemoglobin. An important point to remember is that repeat dosing of sodium nitrite should not be based on the methemoglobin level, because of the inability to measure cyanomethemoglobin. In patients poisoned with cyanide the potential exists for decreasing the oxygen-carrying capacity to critically low levels.[38] The sodium thiosulfate provides substrate for the enzyme rhodenase, which combines thiosulfate and cyanide to form a nontoxic compound thiocyanate. The amyl nitrite capsules should be administered in known or suspected cyanide exposure when intravenous access is not available. The capsule is broken into a gauze and administered for 15–30 seconds every 2–3 minutes. The gauze can be placed between the mask and the mouth if the patient is being ventilated by a bag valve mask or between the oxygen source and the ET tube if the patient is intubated. Once intravenous access is established, sodium nitrite should be administered. The adult dose is 300 mg given intravenously over 2–4 minutes. The antidote package is supplied with 10 ml of 3% solution to be given at 2.5–5.0 ml per minute. The pediatric dose is 6–8 ml/square meter (approximately 0.2 ml/kg or 6 mg/kg) not to exceed 10 ml. The adult dose of sodium thiosulfate is 50 ml of the 25% solution, totaling 12.5 grams. The pediatric dose of sodium thiosulfate is 7 grams/square meter (approximately 1.65 ml/kg) not to exceed 12. grams.[22]

The cyanide antidote package has been administered in one prospective study.[21] The kit was administered based on clinical criteria of a patient who "required intubation, exhibited continued altered mental status or cardiovascular instability or had persistent metabolic acidosis." One patient received treatment when a cyanide level of 3.8 μg/ml was noted 4 hours after admission. All but one patient, who died of anoxic encephalopathy on the sixth day of hospitalization, had good recovery. This paper[21] demonstrates the ability to give the cyanide antidote package without obvious adverse sequelae and shows that methemoglobin levels peaked at a mean of 50 minutes, by which time the carbon monoxide level should already be falling significantly in patients treated with hyperbaric oxygen. The absence of a control group does not allow complete assessment of benefit. Sodium thiosulfate remains the treatment with the least theoretic risk, but the data necessary to make a definitive statement are unavailable.

Several other antidotes are being investigated. Dimethylaminophenol (DMAP) is another methemoglobin-forming agent. Red cell stroma-free methemoglobin, dicobalt EDTA, and hydroxycobalamine are potential antidotes that act by binding cyanide directly, DMAP is currently used in Europe but not approved in the United States. It may have small advantage over nitrites in that it causes less hypotension and induces methemoglobinemia more rapidly. DMAP still has the major disadvantage of decreasing oxygen carrying capacity. Stroma-free methemoglobin has theoretical concerns about obstruction of renal tubules and allergic reactions.[38] Dicobalt EDTA (Kelocyanor) is currently used in Europe but is not approved in the United States. It chelates cyanide and is excreted in the urine. Side effects include allergic reaction, hemodynamic instability, and tachypnea. The most promising new antidote is hydroxycobalamine. This drug is a precursor of vitamin B_{12} and complexes with cyanide to form cyanocobalamine (vitamin B_{12}), which is eliminated by the urine. Hydroxycobalamine is considered an orphan drug by the FDA and is used in Europe.[14,15,35] Side effects include reddish discoloration of the skin, mucous membranes, and urine, increased blood pressure, and anaphylactoid reactions. If hydroxycobalamine proves to have minimal risk, empiric treatment with sodium thiosulfate and hydroxycobalamine may become the treatment of choice.

Management of Methemoglobinemia

Methemoglobinemia in smoke inhalation patients is relatively rare and will rarely require treatment. Treatment of methemoglobinemia involves the administration of methylene blue. This antidote is reduced by the NADPH methemoglobin reductase enzyme and in turn reduces the methemoglobin to normal hemoglobin. Methylene blue is given 1–2 mg/kg over 5 minutes in a 1% solution (10 mg/ml).[19,34] Indications for methylene blue are change in mental status, acidosis, EKG changes, and ischemic chest pain. The presence of subtle mental status changes, tachypnea, and tachycardia thought to be due to hypoxia should be treated. Methemoglobin levels less than 30% may not require treatment, but this depends on the patient's cardiac and respiratory reserve and the presence of preexisting anemia. The main contraindication to methylene blue is G-6-PD deficiency, which predisposes patients to treatment failure and possible hemolysis. In critically ill patients the drug may need to be given prior to testing for G-6-PD deficiency. In the setting of smoke inhalation the treatment of methemoglobinemia has the theoretical risk of releasing cyanide bound to methemoglobinemia. There is no readily available test for cyanomethemoglobin. It will not be detected by routine blood gas analysis. Treatment should be initiated only if the benefit outweighs the risk.

Corticosteroids

Because of the pathophysiology of irritants, the use of corticosteroids to reduce inflammatory response is an obvious consideration. The use of these drugs in smoke inhalation is not clearly established. Patients with asthma or COPD may benefit from corticosteroids, but large surface burns are a contraindication. Prophylactic antibiotics are of no benefit and should only be used when infections are documented.[18]

Disposition and Prognosis

Patients with smoke inhalation should be monitored for at least 4 hours in the emergency department. Admission should be considered for patients with a history of closed space exposure for greater than 10 minutes, arterial PO_2 less than 60 mm Hg, serum bicarbonate less than 15 mmol/L carbonaceous sputum, carboxyhemoglobin levels above 15%, arteriovenous oxygen difference (on 100% oxygen) greater than 100 mm Hg, abnormal chest radiograph, or bronchospasm.[5,25] Admission into intensive care should be considered for patients with hypoxemia, tachypnea, rales, increasing acidosis, or other signs of parenchymal damage. Respiratory failure may develop 36 hours after inhalation of poorly soluble gases.[25]

While the mortality of an isolated smoke inhalation injury is less than 10%, the existence of a smoke inhalation injury in a burn patient can almost quadruple the mortality rate. Other long-term sequelae that must be monitored are subglottic stenosis,[12] bronchiectasis, pulmonary edema (4–9%), bronchopneumonia (3–23%), atelectasis (2.6–11%), pleural effusion (2%), pulmonary embolism (1–2%), and pneumothorax (1–5%).[37] Reactive airway disease can develop after exposure to respiratory mucosal irritants (i.e., phosgene, chlorine, sulfur dioxide) but are more likely to occur after multiple exposure episodes (i.e., firefighters).[28,37] Polyvinylcholoride (PVC) fumes may be particularly responsible for the development of airway hyperresponsiveness, with an incidence following an exposure as high as 18%.[27,28]

REFERENCES

1. Anderson RA, Jarland WA: Fire deaths in the Glasgow area. III. The role of hydrogen cyanide. Med Sci Law 22:35–40, 1982.
2. Baud FJ, Barriot P: Elevated blood cyanide concentrations in victims of smoke inhalation. N Engl J Med 325:1761–1766, 1991.
3. Bresnitz EA. Simple asphyxiants and pulmonary irritants. In Goldfrank LR, et al (eds): Goldfrank's Toxicologic Emergencies. 5th ed. East Norwalk, CT, Appleton & Lange, 1994.
4. Buckley RG, Aks SE: The pulse oximetry gap in carbon monoxide intoxication. Ann Emerg Med 24:252–255, 1994.
5. Clark CJ, Campbell D, Reid WH: Blood carboxyhaemoglobin and cyanide levels in fire survivors. Lancet 2:1332–1335, 1981.
6. Clark CJ, Reid WH, Pollock AJ et al: Role of pulmonary alveolar macrophage activation in acute lung injury after burns and smoke inhalation. Lancet 2:872–874, 1988.
7. Clark WR: Smoke inhalation: Diagnosis and treatment. World J Surg 6:24–29, 1992.
8. Clark WR, Grossman ZD, Ritter-Hrncirik C, et al: Clearance of aerosolized 99Tc-diethylenetriamine pent-acetate before and after smoke inhalation. Chest 94:22–27, 1988.

9. Davis KM, Mutch RW: Wildland fires: Dangers and survival. In Auerbach PS: Wilderness Medicine. 3rd ed. St. Louis, Mosby, 1992.
10. Desai MH, Rutan RL, Herndon DN: Managing smile inhalation injuries. Postgrad Med 8:69–76, 1989.
11. Dinerman N.: Fire hazards (toxic products of combustion). In Rumack BH, Hess AJ, Gelman CR (eds): Poisindex System. Vol 86. Englewood, CO, Micromedex.
12. DiVincenti FC, Pruit, Reckler JM: Inhalation injuries. J Trauma 11:109–117, 1971.
13. Dressler DP: Laboratory background on smoke inhalation. J Trauma 19:913–915, 1979.
14. Forsyth JC, Mueller PD, Becker CE, et al: Hydroxycobalamin as a cyanide antidote: Safety efficacy and pharmacokinetics in heavily smoking normal volunteers. J Toxicol Clin Toxicol 31:277–294, 1993.
15. Hall AH, Rumack BH: Hydroxycobalamin/siduum thiosulfate as a cyanide antidote. J Emerg Med 5:115–121, 1987.
16. Haponik EF: Clinical smoke inhalation injury: Pulmonary effects. Occup Med State Art Rev 8:431–467, 1993.
17. Harwood B, Hall JR: What kills in fires: Smoke inhalation or burns? Fire J 84:29–34, 1989.
18. Heimbach DM, Waeckerle JF: Inhalation injuries. Ann Emerg Med 17:1316–1320, 1988.
19. Howland AH, Mehtylene Blue. In Goldfrank LR, et al (eds): Goldfrank's Toxicologic Emergencies. 5th ed. East Norwalk, CT, Appleton & Lange, 1994.
20. Jones J, McMullen MJ, Doughterty J: Toxic Smoke Inhalation: Cyanide poisoning in fire victims. Am J Emerg Med 5:318–321, 1987.
21. Kirk MA, Gerace R, Kulig KW: Cyanide and methemoglobin kinetics in smoke inhalation victims treated with the cyanide antidote kit. Ann Emerg Med 22:1413–1418, 1993.
22. Kirk MA: Smoke Inhalation. In Goldfrank LR, et al (eds): Goldfrank's Toxicologic Emergencies. 5th ed. East Norwalk, CT, Appleton & Lange, 1994.
23. Koren G, Gorrettson LK, Hill RK, et al: A multicenter, prospective study of fetal outcome following accidental carbon monoxide poisoning in pregnancy. In Koren G: Maternal Fetal Toxicology: A Clinician's Guide. 2nd ed. New York, Marcel Dekker, 1994, pp 253-266.
24. Lindquist P, Rammer L, Sorbo B: The role of hydrogen cyanide and carbon monoxide in fire casualties: A prospective study. Forensic Sci Int 43:9–14, 1989.
25. Liu D, Olson KR: Smoke Inhalation: Contemporary management in critical care. In Hoffman RS, Goldfrank LR (eds): Critical Care Toxicology. New York, Churchill Livingstone, 1991, pp 203–224.
26. Longo LD. The biological effects of carbon monoxide on the pregnant women, fetus, and newborn infant. Am J Obstet Gynecol 129:69–102, 1977.
27. Markowitz J: Self-reported, short and long term respiratory effects among PVC-exposed firefighters. Arch Environ Health 44:30–32, 1989.
28. Moisan TC: Prolonged asthma after smoke inhalation: A report of three cases and a review of previous reports. J Occup Med 33:458–461, 1991.
29. Morikawa T, Yanai E, Nishina T: Toxicity evaluation of fire effluent gases from experimental fires in building. J Fire Sci 5:248–271, 1987.
30. Neiman: The effect of smoke inhalation on pulmonary surfactant. Ann Surg 191:171–181, 1980.
31. Orzel, RA: Toxicological aspects of firesmoke: Polymer pyrolysis and combustion. Occup Med State Art Rev 8:415–29, 1993.
32. Philips AW, Cope O: Burn Therapy: III. Beware the facial burn! Ann Surg 156:759–766, 1962.
33. Piantadosi CA: Carbon monoxide, oxygen transport, and oxygen metabolism. J Hyperbaric Med 2:27–44, 1987.
34. Price D: Methemoglobinemia. In Goldfrank LR, et al (eds): Goldfrank's Toxicologic Emergencies. 5th ed. East Norwalk, CT, Appleton & Lange, 1994, pp 1169–1180.
35. Ruddy, RM: Smoke inhalation injury. Pediatr Clin North Am 41:317–36, 1994.
36. Seger D, Welch L: Carbon monoxide controversies: Neuropsychologic testing, mechanism of toxicity, and hyperbaric oxygen 24:242–248, 1994.
37. Sharar SR, Hudson LD: Toxic gas, fume and smoke inhalation. In Parrillo JE, Bone RC (eds): Critical Care Medicine. St. Louis, Mosby, 1994, pp 849–863.
38. Shusterman D: Clinical smoke inhalation injury: Systemic effects. Occup Med State Art Rev 8:469–503, 1993.
39. Silverman SH, Purdue GF, Hunt JL, Bost RO: Cyanide toxicity in burned patients. J Trauma 28:171–176, 1986.
40. Symington IS: Cyanide exposure in fires. Lancet 91–92, 1978.
41. Thom SR, Taber RL, et al: Delayed neuropsychologic sequelae after carbon monoxide poisoning: Prevention by treatment with hyperbaric oxygen. Ann Emerg Med 25:474–480, 1995.
42. Tibbles PM, Perrotta PL: Treatment of carbon monoxide poisoning: A critical review of human outcome studies comparing normobaric with hyperbaric oxygen. Ann Emerg Med 24:269–276, 1994.
43. Tomaszewski CA: Carbon monoxide. In Goldfrank LR, et al (eds): Goldfrank's Toxicologic Emergencies. 5th ed. East Norwalk, CT, Appleton & Lange, 1994, pp 1199–1214.

44. Tomaszewski CA, Thom SR: Use of hyperbaric oxygen in toxicology. Emerg Med Clin North Am 12:437–459, 1994.
45. Touger M, Gallagher EJ, Tyrell. Relationship between venous and arterial carboxyhemoglobin levels in patients with suspected carbon monoxide poisoning. Ann Emer Med 25:481, 1995.
46. Undersea and hyperbaric medical society: Hyperbaric oxygen therapy, a committee report. UHMS number 30 CR(HBO) 1992.
47. Van Meter KW, Weiss L, Harch PG, et al: Should the pressure be off or on in the use of oxygen in the treatment of carbon monoxide-poisoned patients? Ann Emerg Med 24:283–288, 1994.
48. Vegfors M, Lenmarken C: Carboxyhaemoglobinemia and pulse oximetry. Br J Anaesth 66:625–626, 1991.
49. Weiss LD, Van Meter KW: The applications of hyperbaric oxygen therapy in emergency medicine. Am J Emerg Med 10:558–568, 1992.
50. Wroblewski DA, Bower FC: The significance of facial burns in acute smoke inhalation. Crit Care Med 7:335–338, 1979.
51. Yen D, Tsai J, Wang LM, et al: The clinical experience of acute cyanide poisoning. Am J Emerg Med 13:524–528, 1995.
52. Cyanide Antidote Package, package insert. San Clemente, CA, Pasadena Research Laboratories, 1994.

PAUL A. REICHELT, PhD
KAREN M. CONRAD, RN, MPH, PhD

MUSCULOSKELETAL INJURY:
Ergonomics and Physical Fitness in Firefighters

From the Administrative Nursing and Occupational Health Nursing Program
Public Health Nursing
College of Nursing
The University of Illinois at Chicago
Chicago, Illinois

Reprint requests to:
Paul A. Reichelt, PhD
Associate Professor
Public Health, Mental Health, and Administrative Nursing
College of Nursing
The University of Illinois at Chicago
845 South Damen Avenue
Chicago, IL 60612-7350

The fire service continues to be one of the most hazardous industries in this country. Its rates of work-related injury/illness and fatality exceed those for most industries.[8,36,41,43,56] For example, the incidence of work-related injury in the fire service is reported to be 4.7 times that for private industry, with 41% of firefighters reporting injuries in 1993, a substantial increase from the 36% reporting injuries in 1992.[41] For 1993, the National Fire Protection Association estimated that there were 101,500 fire service injuries that required medical treatment or resulted in at least 1 day of restricted activity,[43] a 3.9% increase from 1992. Recently, lost work hours were reported to be 9.5 times greater in the fire service than in private industry.[41]

Musculoskeletal injuries account for almost half of all line-of-duty injuries among the 1 million firefighters in this country[41,43] and result in lost work time, costly medical claims, and even permanent disability and premature retirement. These injuries are primarily sprains, strains, and muscular pain that most frequently affect the back.[49] Today, it is increasingly common for firefighters to be cross-trained as firefighters and paramedics or emergency medical technicians (EMT), because more than 60% of fire department alarms are for emergency medical services rather than for fire suppression.[41] Therefore, the previously noted statistics reflect injuries and illnesses that result from engagement in fire suppression, rescue, and emergency medical operations. Studies of nonfirefighter paramedics and EMTs report similar statistics for musculoskeletal injuries.[39,61]

The ranking of musculoskeletal injuries as the

primary and most costly injury in the fire service parallels the case for U.S. workers in general.[3,10,24,26,42,58] For adults of working age (18–64 years old), the estimated cost of musculoskeletal conditions was $79 billion in 1988 alone.[51] Over the last few years, the incidence of musculoskeletal injuries has been increasing at the extraordinary rate of about 20% per year.[55] Because musculoskeletal injuries are the primary cause of employee absenteeism and disability, they are of increasing concern to employers, workers, insurance carriers, regulatory agencies, and occupational health professionals.

CONCEPTUAL FRAMEWORK FOR FIRE SERVICE MUSCULOSKELETAL INJURY

The National Institute for Occupational Safety and Health (NIOSH) document on musculoskeletal injuries suggests that preventive strategies consider workplace hazards (such as heavy lifting and repetitive, forceful manual twisting) as well as behavioral or lifestyle factors (such as physical fitness activities).[26] The combination of worksite hazard control and worksite health promotion can clearly enhance the level of workers' health both on and off the job.[19] Unfortunately, there is little theory or research that comprehensively integrates worksite hazard control and health promotion.[18–20] Despite the contrasting nature of these two approaches, they are in fact complementary and can be synergistic.[19]

Taken together, NIOSH's directive and the available research point to the need for a conceptual framework that integrates hazard control and health promotion approaches. Such an integration considers both personal factors, such as physical fitness, and workplace factors, such as biomechanical stressors, as contributors to musculoskeletal injuries. This perspective is sometimes referred to in the literature as *ecological*. For our current program of research, the authors have developed an ecological model of occupational health that builds on the work of several experts.[19,35,50,65] The model posits that desired health outcomes, such as reduced musculoskeletal injury, are determined both by the individual and by factors beyond the individual, in this case the workplace and the external environment in which the fire suppression, rescue, or emergency medical service occurs. The model maintains that the factors are interrelated and that prevention efforts must be directed at both the individual and workplace.

To elaborate the ecological model from the perspective of the fire service, the authors conducted a focus group study in which broad, open-ended questions were asked about what person, workplace, and situational factors fire service participants thought contributed to musculoskeletal injury.[21] No leading questions were asked about specific factors identified in the musculoskeletal injury literature. The purpose of this qualitative study was to help identify factors involved in producing musculoskeletal injuries in firefighters and possibilities for reducing the impact of these factors. The generalizability of these results to the entire fire service population requires further verification.

A series of five focus group sessions was conducted with a total of 11 fire chiefs and 28 firefighters/paramedics from 14 fire departments. Regardless of whether an individual was a fire chief or nonmanagement personnel, exercised or not, or had a compensated time-lost injury, there was agreement on a number of issues. Getting injured as a firefighter is inevitable, even though the injury may not result in time lost from the job. Personal factors related to injury include lifestyle practices (including physical fitness and nutrition), fatigue, commitment to the job, mental preparedness (including common sense), skill level, age, experience, and a sense of team membership.

Workplace factors related to injury include job tasks (such as lifting victims while the firefighter is in an awkward position), equipment, personnel selection, staffing levels, safety training, physical fitness programs, and work schedule. Uncontrollable situational factors include the rescue situation (including working in confined, awkward places), scale of the emergency, unpredictability of the situation, and atmospheric conditions (such as cold, ice, or heat). Our ecological model, as elaborated from the focus group data, is shown in Figure 1.

Participants differed in some of their perceptions. Although participants agreed that physical fitness is important, they did not agree on the importance of fitness standards or on mandatory fitness programs. They also differed in how much weight they attributed to each factor. The chiefs considered personal factors the linchpin of musculoskeletal injuries, attributing an average of 50% of injuries to personal factors. In contrast, firefighters attributed an average of 26% of injuries to personal factors. Fire chiefs attributed 28% of injuries to uncontrollable factors, whereas firefighters attributed 40% of injuries due to these factors. Thus, both chiefs and firefighters believed that uncontrollable factors account for some, but not all, musculoskeletal injury. They believed that injuries could be reduced by addressing selected personal and workplace factors. A comparison of chiefs' and firefighters' perceptions of the relative contribution of each factor to musculoskeletal injuries appears in Figure 2.

Thus, the focus group study suggests that both the workplace and person present opportunities for preventive intervention. The major workplace factors include biomechanical stressors created by the performance of firefighter/paramedic tasks and the interaction with fire service equipment. The individual differences or personal factors include anthropometric and physical capacity measures. Situational factors at the scene of the emergency comprise the external environment of interest. The combined contribution of personal and workplace factors to musculoskeletal injury suggested by these groups is significant enough that it seems prudent to pursue development of integrated preventive measures. It is confirming that the groups listed many of the same personal and workplace factors identified in the musculoskeletal injury literature despite no probing in this regard by the focus group moderator. However, the study went beyond generating a list of factors. The information gathered provided insight into why the problems occur and the context in which they occur. This qualitative data, in conjunction with the existing literature, laid the groundwork for developing the conceptual model that can be tested in larger studies. In the remainder of this chapter, we focus our discussion on two key factors identified by the groups as contributing to the production of fire service musculoskeletal injuries: ergonomics and physical fitness.

RESEARCH ON MUSCULOSKELETAL INJURIES IN OTHER OCCUPATIONS: ERGONOMICS AND PHYSICAL FITNESS

Although the literature on firefighter factors that contribute to musculoskeletal injury is sparse, with an emphasis on individual fitness, there is a considerable amount of epidemiologic research not involving firefighters that delineates the contribution of workplace and personal factors. Workplace risk factors for back injuries include lifting heavy objects, twisting, work postures, sudden maximal effort, stretching and reaching, heavy manual labor, cumulative load exposure, as well as working more than 40 hours per week, equipment design, worker selection, safety training, medical services, and driving and vibration (as may occur in fire trucks and ambulances traveling at high speeds).[4,10–12,17,24,26,44,45,52,59,60,64] Many of these factors are ergonomic in nature.

Related to ergonomics, a recent study examined the link between trunk motions, workplace factors, and the incidence of occupational low back disorders.[48] The purpose of this study was to determine whether dynamic trunk motions, along with other risk factors, could better describe the risk of low back disorder in repetitive manual materials handling. The "Lumbar Motion Monitor" was used to document the three-dimen-

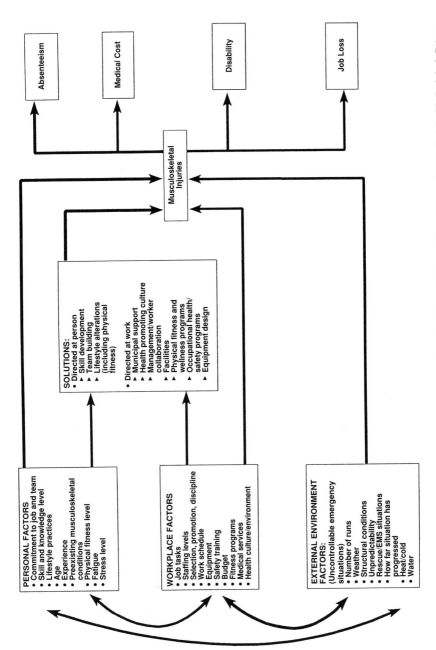

FIGURE 1. Ecological model of fire service musculoskeletal injuries. (Adapted from Conrad KM, Balch GI, Reichelt PA, et al: Musculoskeletal injuries in the fire service: Views from a focus group study. AAOHN J 42:572–581, 1994.)

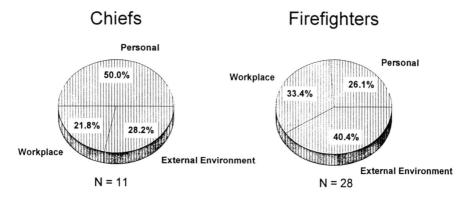

FIGURE 2. Chiefs' and firefighters' perceptions of the relative contribution of personal, workplace, and uncontrollable external environment factors to musculoskeletal injuries. (Adapted from Conrad KM, Balch GI, Reichelt PA, et al: Musculoskeletal injuries in the fire service: Views from a focus group study. AAOHN J 42:572–581, 1994.)

sional components of trunk motion in the work environment. In reporting a five-factor model, these investigators showed that the maximal amount of forward bending, the peak side-bending velocity, and the average twisting velocity were the trunk motion components that best discriminated between groupings of jobs with historically a high or low incidence of low back disorders. The remaining two factors in the model were the rate at which lifts are performed per hour and the maximum moment imposed on the spine by the external load (object lifted). Although the rate of lifting in the fire service may not be high, the loads lifted (e.g., victims) are often quite heavy. The emergency nature of the job can result in rapid and extreme motions of the torso as firefighting and paramedic tasks are performed quickly.

Personal factors identified in the nonfirefighter literature include age, gender, anthropometry, posture, spine mobility, physical fitness, body strength, diet, stress, fatigue, preexisting medical treatment for both back and nonback conditions, job experience, job attitude, job dissatisfaction, and poor employee performance rating.[3,4,9,10,12,24,26,44,64] This literature notes the role of physical fitness as well as individual ergonomic factors.

As previously noted, these results are consistent with the results of the focus group study in which the authors elaborated their ecological model (Fig. 1).[21] The firefighters/paramedics expressed concern regarding both their physical fitness and ergonomic factors such as the lifting, stretching and bending, and twisting associated with rescue operations and emergency medical service provision. In sum, the available literature points to the importance of considering workplace factors in addition to personal factors, as has previously been emphasized in characterizing musculoskeletal injuries among fire service personnel.

FIREFIGHTING JOB TASKS: ERGONOMICS AND PHYSICAL FITNESS ON THE JOB

A few studies have examined firefighting tasks in relation to the physical requirements needed to perform the job.[25,27,28,30,33,46,47,56] In their study, Dotson et al. quantified the physiologic and physical requirements of the structural firefighter.[30] For this seminal study of 100 career firefighters, the Fire Training Offices of the Metropolitan Council of Governments in Washington, DC, nominated a representative list of frequent and critical firefighting tasks. Five tasks from the nominated list were selected for use in the study and have remained essentially the same since (stair/ladder climb, hose advance, simulated forcible entry/chopping, hoist evolution/standpipe-hose load pull, and simulated victim rescue/dummy drag). Performance of the five tasks was timed, and correlation analysis identified key variables relating physical performance and the simulated firefighting tasks. The two primary factors operating in the simulated tasks were high aerobic energy involvement and resistance to fatigue. The study concluded that high muscular strength and endurance as well as near-maximal aerobic capacity were required to complete the simulated tasks representing fire suppression job performance.

In another job task analysis, Gledhill and Jamnik characterized the physical demands of Canadian firefighters.[33] Like the study by Dotson et al.,[30] the focus in this study was on quantifying the physical requirements of the job, such as aerobic and anaerobic energy requirements, muscular strength and endurance, and flexibility. These studies concentrated on fire suppression tasks.

Only one firefighter study was found that quantified biomechanical stressors,[47] and, as in the other job task analyses, the task analysis related to fire suppression. Lusa and colleagues evaluated the biomechanical load factors in the simulated clearing of passages through the ceiling with the use of a power saw in younger (32 ± 2 years) and older (47 ± 5 years) Scandinavian firefighters.[47] The authors concluded that lifting a power saw produced a high load on the musculoskeletal system and that the load was not influenced by age. The results highlighted the importance of muscular strength and proper work techniques in the use of heavy manual tools. No firefighter job task analysis studies were found that centered on medical emergency rescue tasks.

These fire suppression task analyses are consistent with those of Doolittle and Kaiyala, who concluded that there was an inverse association between strength and musculoskeletal injury.[29] This was the only firefighter study reviewed that attempted to control for risk of exposure to musculoskeletal injury. In their study of multiple fire stations, exposure was defined as the number of hours spent on the fire scene per fire station. Although this is an admittedly gross measure of exposure, it does represent acknowledgment that exposure needs to be quantified in studies of fire service musculoskeletal injuries.

In a similar vein, a large prospective study of Los Angeles County firefighters related preemployment physical measurements to subsequent work-related back injuries.[62] This study concluded that trunk flexibility was the strongest protective factor against the occurrence of work-related back injuries. Given these findings, it is interesting that there are no federal mandatory standards on physical fitness for the fire service, and overall, this worker group is no more physically fit than the rest of the North American population.[49]

PHYSICAL FITNESS AND FATIGUE

Our conceptual model shows that in addition to ergonomic factors at the personal and workplace levels, physical fitness factors at these two levels are another important potential point of intervention. And, in fact, physical fitness and fatigue are two variables that research suggests are important determinants of musculoskeletal injury.[5,10] Not only are these two variables each hypothesized to influence musculoskeletal injury directly, but it is likely that physical fitness mitigates against physical and mental fatigue, acting as a mediating variable in the prediction of musculoskeletal injury.[37] Thus, compared to a less physically fit worker, a physically fit worker is expected to perform better on the complex mental job tasks that currently face firefighters/paramedics and is expected to have a faster recovery rate from mental fatigue.[31] In terms of physical fatigue, a physically fit firefighter/paramedic is expected to have greater cardiovascular and muscular endurance in performing physical fire service tasks than is a less-fit worker.[2]

Although the linkage between fitness and fatigue is intuitively appealing, the research is underdeveloped.[63] Because the fire service provides around-the-clock services delivered by personnel on duty in 24-hour shifts, fatigue would appear to be an important personal factor. However, there is only one reported empirical study that examined the link between physical fitness and reaction time (an indicator of fatigue) in the fire service[6] and no studies on fatigue and musculoskeletal injury. Our review of published physical fitness intervention studies in the fire service showed that with few exceptions, the primary focus has been on cardiovascular risk reduction rather than musculoskeletal injury risk reduction.

PHYSICAL FITNESS INTERVENTIONS FOR FIREFIGHTERS

As previously noted, an inverse relationship between physical fitness and the likelihood of musculoskeletal injuries in firefighters has been suggested by a number of investigators.[16,21,38,49] In a focus group study by Conrad et al., chiefs and firefighters believed that though musculoskeletal injuries are an inevitable consequence of the job, being physically fit is one thing firefighters can do to minimize the risk and severity of musculoskeletal injuries.[21] Although still not common in the fire service, firefighter physical fitness programs have been developed primarily over the last 15 years to combat the risk for musculoskeletal and traumatic injuries and for heart disease. Unfortunately, the programs have not been convincingly evaluated for their effectiveness.[36]

This section of the chapter integrates findings from the firefighter physical fitness intervention studies conducted to date and summarizes the state of knowledge on the topic. A comprehensive literature search identified 11 physical fitness intervention studies for firefighters that reported longitudinal data analysis and that were published in scientific journals. The studies spanned the years 1979–1991.

Eight of the 11 studies were conducted in the United States.[1,6,7,16,32,34,38,53] The others took place in England,[14] Canada,[54] and the Auckland region in New Zealand.[40] Ten of the studies were conducted in urban fire departments.[1,6,7,14,16,34,38,40,53,54] Sample sizes ranged from 24 in the study by Green and Crouse[34] to 2,856 in Barnard's and Anthony's study.[7] In most cases, the sample size was less than 100.[1,6,14,32,34,53]

The studies showed considerable variability in the design of the research, type of fitness program, and outcomes assessed. Reid's and Morgan's early study was the only one to use an experimental design with firefighters randomly assigned to one of three increasingly comprehensive treatment groups.[54] The remaining studies used a quasi-experimental design. Six of the quasi-experimental studies used a one-group pretest-posttest design with anywhere from one to five posttest measurement points.[6,7,16,32,34,40] The remaining four quasi-experimental studies used a pretest-posttest nonequivalent-group design.[1,14,38,53]

Intervention Program Components

Four of the programs were mandatory,[7,16,34,38] six were voluntary,[1,14,32,40,53,54] and one was a voluntary program in year 1 of the study and mandatory in year 2.[6] Only four of the studies did not provide individual exercise prescriptions for program participants that were based on pretest screening results.[7,14,38,40]

All components of physical fitness were addressed by this group of studies, but not necessarily in the same program: cardiorespiratory endurance, body composition, muscular strength and endurance, and flexibility. Generally, standardized measurement protocols were in place. An argument can be made that all of these components of physical fitness can help reduce the firefighter's risk for musculoskeletal injury.[7,16,38] The hypothesized effect may be direct or indirect, as previously discussed.

A variety of exercise modes were used, including walking, jogging, running, cycling, rowing, stair climbing, skipping rope, weight lifting, calisthenic circuits, and stretching. Typically, the firefighters worked out two to three times per week on shift days at the station, with each exercise session lasting around 30-60 minutes. There was a considerable range in the length of time that the programs were studied—from 11 weeks in the study by Brown et al.[14] to 9 years in the work by Cady et al.[16] Five of the programs were studied beyond 1 year.[6,7,16,34,40]

Intervention Program Outcomes

For the most part, the studies examined physiologic outcome measures. The studies reported positive outcomes, including improved VO_{2max}, decreased resting and recovery heart rates, decreased percentage of body fat, increased muscular strength and endurance, increased flexibility, and, in the Bahrke study,[6] decreased reaction time. In most cases, the results were statistically significant, with effects tested primarily using t-tests or analysis of variance procedures.

A few investigators examined the effect of their programs on absenteeism, medical claims and costs, and injuries. In two cases, musculoskeletal injuries, in particular, were examined. For a 2-year post-intervention period, Hilyer et al.'s analysis of a flexibility program found average absenteeism costs to be lower for the experimental group than for the comparison group ($950 versus $2,838, $p=0.026$).[38] Medical costs were also less in the experimental group, but results were not statistically significant. There were fewer strains, sprains, and muscle tears in the experimental group than in the comparison group, with the rates of injury being 19.1/100 in the experimental group and 23.9/100 in the comparison group (no p values given). Bahrke, who compared data from 60 firefighters for 1 year before the program with 1 year after the program, concluded that no tangible benefits were evident for absenteeism, medical claims, or total medical costs.[6] An inspection of the numbers shows no pattern to the data. Finally, Cady et al. conducted a number of separate analyses, not restricting them to the 188 firefighters in the longitudinal study. In a cross-sectional analysis, they found level of physical fitness, operationalized as strength, flexibility, and physical work capacity, to be inversely related to back injuries, as well as back injury costs and total injury costs.[16] Over the 9-year study period, the number of disabling injuries declined and the worker compensation costs decreased by 25%.

Intervention Program Conclusions

Taken as a whole, the handful of physical fitness intervention studies in firefighters suggest the potential for formalized fitness programs to improve the physical fitness of firefighters in terms of cardiorespiratory endurance, body composition, muscular strength and endurance, and flexibility. Although examined in only three studies, results also suggested that physical fitness mitigates against injuries, including those that are musculoskeletal in nature, absenteeism, and medical claims and costs. These findings are consistent with other worksite fitness literature.[57]

Although beyond the scope of this brief overview, questions to the validity of this conclusion of a program effect exist (i.e., selection bias of volunteers, local history, small sample size, inadequate statistical power, compensatory rivalry, using means to compare groups when the data are skewed, etc.) and need to be considered.[22,23] Both voluntary and mandatory programs may produce positive outcomes, as Bahrke noted, but the mandatory program phase of his study showed more positive effects than the voluntary program phase.[6] Whether a program was voluntary or mandatory, allowing the firefighters to exercise during work hours undoubtedly enhanced adherence rates. Exercising during working hours likely promoted social support to exercise. None of the studies addressed the issues of adherence or program dropout rates, an important element to consider when interpreting the validity of the results.[13] Providing firefighters with an individualized exercise prescription that tailored the program to their fitness needs may also have prompted positive fitness changes. Most programs were comprehensive in that they addressed multiple fitness components simultaneously. Such a comprehensive approach makes sense when the goal is to reduce musculoskeletal injury.

LOOKING TOWARD THE FUTURE OF FIRE SERVICE OCCUPATIONAL HEALTH

Clearly, all the pieces to the research puzzle are not in place regarding the relationships among ergonomics, fitness, and musculoskeletal injury in the fire service. However, some evidence suggests that physical fitness functions as predicted in reducing the risk for musculoskeletal injury. This positive finding makes the study of the relationship between fitness and fatigue along with ergonomic factors important for understanding how to reduce musculoskeletal injuries in the fire service. Figure 3, which is derived from our ecological model, highlights the importance of considering both personal and workplace level factors.

The reason for emphasizing personal and workplace factors as points of intervention while deemphasizing external environment factors is specific to the fire service and is not meant to ignore the importance of the environment in the production of musculoskeletal injuries. Although the external environment is generally the most difficult aspect for any organization or occupation to control, this is especially true of the fire service where the external environment factors are basically uncontrollable emergency situations. Other than such activities as working to reduce the number of false alarm runs through public education and working with other public agencies to achieve strict enforcement of the fire safety and building codes, there is little the fire service can do to impact many of the important environment factors, such as the weather or age of the buildings in their area. Therefore, personal and workplace factors represent the better points of entry for occupational health interventions designed to benefit fire service personnel both as individuals and as a collective occupational group.

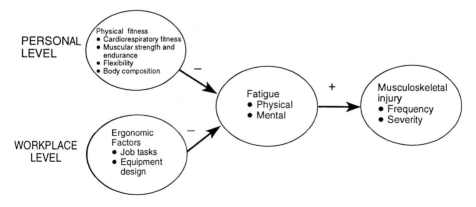

FIGURE 3. Two-level model for controlling fire service musculoskeletal injuries.

Several avenues of investigation show promise for reducing musculoskeletal injuries in the fire service. First of all, to recommend effective and efficient ergonomic interventions, it is necessary to have a better understanding of the biomechanical parameters involved in rescue and emergency medical service (EMS) operations that now represent the majority of fire service alarms. The authors and their research team from the Fire Service Occupational Safety and Health Project at the University of Illinois at Chicago are currently funded by NIOSH to conduct an ergonomic task analysis to identify the rescue and EMS job tasks and to quantify the parameters of the tasks that expose firefighter/paramedics to biomechanical stressors. The knowledge gained from this project will lay the groundwork for integrated intervention studies aimed at altering manipulable person and workplace factors that promote musculoskeletal injuries among fire service personnel.

Similarly, additional work is required to evaluate further the potentially effective and efficient fitness interventions using research methods that maximize the interpretability and utility of the data generated. In addition to looking at first-level outcomes such as participants' levels of cardiorespiratory endurance, body composition, muscular strength and endurance, and flexibility, evaluations of fitness interventions need also to include a comprehensive look at important second-level outcomes, such as musculoskeletal injury rate, absenteeism, disability, medical costs, and workers' compensation insurance costs. The utility of information produced will also be increased if the evaluations are longitudinal and ensure that the fitness intervention groups exercise for a reasonable training period, that research controls are used, and that the evaluations are based in fire service settings where the occupational health problems actually occur and where the preventive strategies need to be implemented. With funding from the National Institute of Nursing Research, the authors are expanding their program of fire service research to begin this needed investigation of physical fitness and fatigue.

SUMMARY

The authors posit that the problem of work-related musculoskeletal injuries in the fire service can best be dealt with by using a conceptual framework that recognizes the interplay between personal and workplace factors and their combined contributions to musculoskeletal injuries. Such an integration should consider both ergonomic and physical fitness factors at both the personal and workplace levels as contributors to musculoskeletal injuries. It is our hope that the ecological model of fire service musculoskeletal injuries we presented in Figure 1 will serve as a beginning step toward the needed integration of these two valuable occupational health approaches. The philosophy guiding our conceptual framework is simply that if occupational health risks to fire service personnel are to be reduced, then hazard control and health promotion approaches should be integrated.Such integration requires that we understand both the workers' and management's views in order to develop programs that are acceptable to both of these major stakeholders. This should be an attainable goal because, as Conrad et al. demonstrated, there are many similar perceptions held by chiefs and firefighters that provide common ground for working collaboratively on solutions to reduce the incidence of musculoskeletal injuries.[21]

REFERENCES

1. Adams TD, Yanowitz FG, Chandler S, et al: A study to evaluate and promote total fitness among fire fighters. J Sports Med 26:337–345, 1986.
2. American College of Sports Medicine: Guidelines for Exercise Testing and Prescription. Philadelphia, Lea & Febiger, 1991.
3. Andersson GBJ: Epidemiologic aspects on low-back pain in industry. Spine 6:53–60, 1981.

4. Andersson GBJ: The epidemiology of spinal disorders. In Frymoyer JW (ed): The Adult Spine: Principles and Practice. New York, Raven Press, 1991, pp 107–146.
5. Andersson GBJ, Pope MH, Frymoyer JW, et al: Epidemiology and cost. In Pope MH, Andersson GBJ, Frymoyer JW, et al. (eds): Occupational Low Back Pain: Assessment, Treatment, and Prevention. St. Louis, Mosby, 1991, pp 95–113.
6. Bahrke MS: Voluntary and mandatory fitness programs for fire fighters. Physician Sports Med 10(8):126–132, 1982.
7. Barnard RJ, Anthony DF: Effect of health maintenance programs on Los Angeles City firefighters. J Occup Med 22:667–669, 1980.
8. Ben-Ezra V, Verstraete R: Stair climbing: An alternative exercise modality for fire fighters. J Occup Med 30:103–105, 1988.
9. Bigos SJ, Battie MC, Spengler DM, et al: A prospective study of work perceptions and psycho-social factors affecting the report of back injury. Spine 16:1–6, 1990.
10. Bigos SJ, Battie MC, Spengler DM, et al: A longitudinal prospective study of industrial back injury reporting. Clin Orthop 279:21–34, 1992.
11. Bigos SJ, Spengler DM, Martin NA, et al: Back injuries in industry: A retrospective study: II. Injury factors. Spine 11:224–251, 1986.
12. Bigos SJ, Spengler DM, Martin NA, et al: Back injuries in industry: A retrospective study: III. Employee-related factors. Spine 11:252–256, 1986.
13. Blue CL, Conrad KM: Adherence to worksite exercise programs: An integrative review of recent research. AAOHN J 43:76–86, 1995.
14. Brown A, Cotes JE, Mortimore IL, Reed JW: An exercise training programme for firemen. Ergonomics 25:793–800, 1982.
15. Cady LD, Bischoff DP, O'Connell ER, et al: Strength and fitness and subsequent back injuries in fire fighters. J Occup Med 21:269–272, 1979.
16. Cady LD, Thomas PC, Karwasky RJ: Program for increasing health and physical fitness of fire fighters. J Occup Med 27:110–114, 1985.
17. Clemmer DI, Mohr DL, Mercer DJ: Low-back injuries in a heavy industry: 1. Worker and workplace factors. Spine 16:824–830, 1991.
18. Cohen A: Evolving efforts at NIOSH to tie worksite health promotion with worksite hazard protection activities. Cincinnati, National Institute for Occupational Safety and Health, 1988.
19. Cohen A: Perspectives on self-protective behaviors and work place hazard. In Weinstein NE (ed): Taking Care: Understanding and Encouraging Self-protective Behavior. New York, Cambridge University Press, 1989.
20. Cohen A, Murphy L: Indicators of health promotion behaviors in the workplace. In Kar SB (ed): Health Promotion Indicators and Actions. New York, Springer, 1989, pp 249–270.
21. Conrad KM, Balch GI, Reichelt PA, et al: Musculoskeletal injuries in the fire service: Views from a focus group study. AAOHN J 42:572–581, 1994.
22. Conrad KJ, Conrad KM: Reassessing validity threats in experiments: Focus on construct validity. In Conrad KJ (ed): Critically Evaluating the Role of Experiments. [New Directions in Program Evaluation Series, Vol 63.] 1994, pp 5–25.
23. Conrad KM, Conrad KJ, Walcott-McQuigg J: Threats to internal validity in worksite health promotion program research: Common problems and possible solutions. Am J Health Promot 6:112–122, 1991.
24. Daltroy LH, Larson MG, Wright EA, et al: A case-control study of risk factors for industrial low back injury: Implications for primary and secondary prevention programs. Am J Ind Med 20:505–515, 1991.
25. Davis PO, Dotson CO, Santa Maria DL: Relationship between simulated fire fighting tasks and physical performance measures. Med Sci Sports Exer 14:65–71, 1982.
26. Department of Health and Human Services: Proposed national strategy for the prevention of musculoskeletal injuries. Washington, DC, National Institute for Occupational Safety and Health, 1986, DHHS publication 89-129.
27. Doolittle TL: Validation of physical requirements for firefighters [technical report prepared for the Seattle Fire Department and Office of Management and Budget], 1979.
28. Doolittle TL: Validation of a firefighter physical ability test [technical report for the City of Tacoma, WA], 1989.
29. Doolittle TL, Kaiyala K: Strength and musculoskeletal injuries of firefighters. Presented at Annual Conference of the Human Factors Association of Canada, Richmond, British Columbia, 1986, pp 49–52.
30. Dotson CO, Santa Maria DL, Davis PO, et al: Development of a job-related physical performance examination for fire fighters. College Park, MD, University of Maryland, 1977.
31. Falkenberg LE: Employee fitness programs: Their impact on the employee and the organization. Acad Manage Rev 12:511–522, 1987.
32. Faria IE, Faria EW: Effect of exercise on blood lipid constituents and aerobic capacity of fire fighters. J Sports Med Phys Fitness 31:75–81, 1991.

33. Gledhill N, Jamnik VK: Characterization of the physical demands of firefighting. Can J Sports Sci 17:207–213, 1992.
34. Green JS, Crouse SF: Mandatory exercise and heart disease risk in fire fighters. Int Arch Occup Environ Health 63:51–55, 1991.
35. Green LW, Kreuter MW: Health Promotion Planning: An Educational and Environmental Approach. Mountain View, CA, Mayfield, 1991.
36. Guidotti TL, Clough VM: Occupational health concerns of firefighting. Annu Rev Public Health 13:151–171, 1992.
37. Harma M: Individual differences in tolerance to shiftwork: A review, Ergonomics 36:101–109, 1993.
38. Hilyer JC, Brown KC, Sirles AT, et al: A flexibility intervention to reduce the incidence and severity of joint injuries among municipal fire fighters. J Occup Med 32:631–637, 1990.
39. Hogya PT, Ellis L: Evaluation of the injury profile of personnel in a busy urban EMS system. Am J Emerg Med 8:308–311, 1990.
40. Hopkins WG, Loughnan Werry JM: The effect of a health development programme on Auckland firefighters. N Z J Health Phys Educ Rec 24:23–24, 1991.
41. International Association of Fire Fighters: 1993 Death and Injury Survey. Washington, DC, International Association of Fire Fighters, 1994.
42. Jacknow D, McCunney RJ, Jofe M: Musculoskeletal disorders. In McCunney RJ (ed): Handbook of Occupational Medicine. Boston, Little, Brown & Co. 1988, pp 106–129.
43. Karter MJ, LeBlanc PR: U.S. Fire fighter injuries—1993. NFPA J 87(7):57–67, 1994.
44. Kelsey JL, Golden AL: Occupational and workplace factors associated with low back pain. Occup Med State Art Rev 3:7–16, 1988.
45. Kumar S: Cumulative load as a risk factor for back pain. Spine 15:1311–1316, 1990.
46. Lemon PW, Hermiston RT: The human energy cost of fire fighting. J Occup Med 19:558–562, 1977.
47. Lusa S, Louhevaara V, Smolander J, et al: Biomechanical evaluation of heavy tool-handling in two age groups of firemen. Ergonomics 34:1429–1432, 1991.
48. Marras WS, Lavender SA, Leurgans SE, et al: The role of dynamic three-dimensional trunk motion in occupationally-related low back disorders: The effects of workplace factors, trunk position, and trunk motion characteristics on risk of injury. Spine 5:617–628, 1993.
49. Matticks CA, Westwater JJ, Himel HN, et al: Health risks to fire fighters. J Burn Care Rehabil 13:223–235, 1992.
50. McLeroy KR, Bibeau D, Steckler A, et al: An ecological perspective on health promotion programs. Health Educ Q 15:351–377, 1988.
51. Praemer MA, Furner S, Rice DP: Musculoskeletal Conditions in the United States. Park Ridge, IL, American Academy of Orthopaedic Surgeons, 1992.
52. Punnett L, Fine LJ, Keyserling WM, et al: Back disorders and non-neutral trunk postures of automobile assembly workers. Scand J Work Environ Health 17:337–346, 1991.
53. Puterbaugh JS, Lawyer CH: Cardiovascular effects of an exercise program: A controlled study among firemen. J Occup Med 25:581–586, 1983.
54. Reid EL, Morgan RW: Exercise prescription: A clinical trial. Am J Public Health 69:591–595, 1979.
55. Rempel D: Ergonomics: Prevention of work-related musculoskeletal disorders. West J Med 156:409–410, 1992.
56. Romet TT, Frim J: Physiological responses to fire fighting activities. Eur J Appl Phys 56:633–638, 1987.
57. Shephard RJ: A critical analysis of work-site fitness programs and their postulated economic benefits. Med Sci Sports Exerc 24:354–370, 1992.
58. Snook SH: The cost of back pain in industry. Occup Med State Art Rev 3:7–16, 1988.
59. Snook SH: Approaches to the control of back pain in industry: Job design, job placement, and education/training. Occup Med State Art Rev 3:45–59, 1988.
60. Snook SH, Fine LJ, Silverstein BA: Musculsokeletal disorders. In Levy BS, Wegman DH (eds): Occupational Health: Recognizing and Preventing Work-Related Disease. Boston, Little, Brown & Co., 1988, pp 345–370.
61. Stilwell JA, Stilwell PJ: Sickness absence in an ambulance service. J Soc Occup Med 34:96–99, 1984.
62. Sullivan C, Yuen J: Predictors of work-related back injuries among fire fighters. In Proceedings of the Fifth International Symposium on Epidemiology in Occupational Health, 1986, p 165.
63. Thomas JR, Landers DM, Salazer W, et al: Exercise and cognitive function. In Bouchard C, Shephard RJ, Stephens T (eds): Physical activity, fitness, and health. Champaign, IL, Human Kinetics, 1994.
64. Venning PJ, Walter SD, Stitt LW: Personal and job-related factors as determinants of incidence of back injuries among nursing personnel. J Occup Med 29:820–825, 1987.
65. Winnett RA, King AC, Altman DG: Health psychology and public health. New York, Pergamon, 1989.

VIRGINIA M. WEAVER, MD, MPH
SHARON DOYLE ARNDT, MPH

COMMUNICABLE DISEASE AND FIREFIGHTERS

From the Division of Occupational Health
Johns Hopkins University School of Hygiene and Public Health
Baltimore, Maryland (VMW)
and
Department of Occupational Health and Safety
International Association of Fire Fighters
Washington, DC (SDA)

Reprint requests to:
Virginia M. Weaver, MD,MPH
Division of Occupational Health
Johns Hopkins University School of Hygiene and Public Health
615 North Wolfe Street
Room 7041
Baltimore, MD 21205

In the past decade, infectious diseases have become a prominent concern in the fire service. Bloodborne pathogens and tuberculosis are the most significant biological hazards; however, firefighters have the potential for exposure to many other communicable diseases. Fire service responsibilities are no longer limited to the traditional roles of extinguishment, search, and rescue. Increasingly, firefighters are cross-trained in emergency medical response roles. A recent national survey of U.S. and Canadian paid fire departments found that 90% of all firefighters provide some level of medical care in the community.[3] On average, 77% of the fire departments in U.S. and Canadian cities with populations of 1 million or more provide first-responder services, 80% provide basic life support, and 50% provide advanced life support.[3]

This increased responsibility places firefighters at risk for many of the communicable diseases that are well-known occupational hazards for health care workers in medical facilities. A study of infectious disease exposures in 650 dual-trained firefighter/emergency medical technicians from 1988–1989 found an incidence rate of 1.3 exposures to blood or other potentially infectious materials per 1,000 emergency medical service (EMS) calls.[49] During that same period, approximately 1 in every 1,000 EMS calls resulted in exposure to a respiratory pathogen.

The working environment of firefighters in both fire suppression and medical response is an important contributing factor to their increased occupational risk for infectious disease. Due to varying work sites, engineering controls, such as those used for tuberculosis in hospitals, are not possible. Working space is often limited, resulting in worker crowding, which may increase the potential for needlestick injury. Space limitations also increase the opportunity for exposure to respiratory pathogens, such as in the back of the ambulance.

To prevent morbidity and, in some cases, mortality from such exposures, it is important to identify the occupational biologic hazards for firefighters and ensure that appropriate prevention strategies are in place. This chapter discusses the pertinent infectious diseases and methods to reduce their incidence in the fire service. Applicable legislation and guidelines are reviewed. In addition, resources specific to the fire service are identified.

BLOODBORNE PATHOGENS

Firefighters involved in emergency medical response are exposed to bloodborne pathogens through contact with blood and other potentially infectious materials, either by medical procedures, such as intravenous line insertion, or by the nature of the patient's injury. Emergency response often involves care of patient populations who have an increased prevalence of infection with one or more bloodborne pathogens. Basic firefighting activities may also involve contact with potentially infectious materials in emergency rescue situations of victims who have been injured and are actively bleeding or severely burned. The firefighter's intact skin barrier may be breached by abrasions and lacerations occurring in fire suppression or rescue activities (e.g., injury from broken glass). In addition, in critical situations, the emergency provider may act in haste with disregard for his or her own safety. All of these factors combine to place the firefighter at increased risk for contracting a bloodborne disease.

The Occupational Safety and Health Administration (OSHA) included firefighters in the estimated population covered by the Bloodborne Pathogens Standard.[45] OSHA survey responses for the standard indicated that 98% of emergency medical technicians (EMTs) and 80% of firefighters are potentially exposed.

In the standard, occupational exposure is defined as the opportunity for "skin, eye, mucous membrane, or parenteral contact with blood or other potentially infectious materials."[45] A variety of body fluids are considered potentially infectious, such as semen, vaginal secretions, cerebrospinal fluid, synovial fluid, and any fluid with visible blood contamination. Fluids in situations in which the detection of blood contamination is difficult are also included (e.g., emergency response settings with inadequate lighting). Although the transmission potential for biologic materials other than blood, semen, and vaginal secretions is unknown, other body fluids are included due to theoretical concerns.

Hepatitis B

Hepatitis B virus (HBV) is the most infectious occupational bloodborne pathogen. It is estimated that approximately 1 million persons are chronic HBV carriers in the United States.[6] These individuals, along with an estimated 300,000 acutely infected patients annually, constitute a substantial source for occupational exposure. The patient population served by EMS personnel has a higher HBV infection rate than that of the general population. A multicenter survey of emergency department patients in Portland, Oregon, found that 12% of 444 blood samples were positive for the antibody to hepatitis B core antigen (anti-HBc) and 0.6% were positive for the hepatitis B surface antigen (HBsAg).[24] Five percent seropositivity was noted in a study of 612 emergency department patients in Baltimore.[25] Depending on the population, the prevalence of HBsAg positivity can range from 0.2 to 20% in high-risk groups, such as clients in institutions for the developmentally disabled.[6]

Numerous studies have documented an increased prevalence of HBV serologic markers in health care workers and, more specifically, in the fire service compared to the general population. A seroprevalence study of 338 Houston firefighters assigned to EMS duties revealed that 13% were positive for one or more of the three HBV tests obtained (HBsAg, anti-HBs, and anti-HBc).[48] Anti-HBs and anti-HBc prevalence increased with years of occupational exposure; this finding was statistically significant and unrelated to age. In comparison, the HBV prevalence in Houston blood donors at that time was 5%.[48] Kunches et al. also found an elevated prevalence of the same HBV markers (18%) in their survey of 87 EMS personnel in Boston.[29] A study of Seattle paramedics revealed that 22% were seropositive for anti-HBs or anti-HBc, which was five times higher than that in the general population in that area.[56]

Conversion rates in nonimmune persons who experience needlestick injuries from HBV-infected sources are substantial, ranging from 2% to as high as 40% if the source patient is HBeAg-positive.[19] Although household transmission of HBV can occur, percutaneous blood exposure is the main occupational route. Infected individuals are at risk for development of chronic infection, which may result in cirrhosis or hepatocellular carcinoma. The Centers for Disease Control and Prevention (CDC) estimated that 5,100 health care workers acquired an occupational HBV infection in 1991.[53] Based on historical experience, the CDC estimates that 5% were hospitalized; 0.1% died from hepatic failure; and 10% became chronic carriers, of whom 20% will ultimately die of cirrhosis or hepatocellular carcinoma. These numbers, although still unacceptably high given vaccine availability, are substantially decreased from the CDC estimate of 12,000 annual infections in health care workers in the mid-1980s, before vaccination was widely adopted.[4]

Hepatitis C

The recent introduction of serologic tests for hepatitis C virus (HCV) has greatly increased knowledge regarding the epidemiology of this virus and its potential as an occupational health risk. An enzyme-linked immunosorbent assay (ELISA) screen with recombinant immunoblot assay (RIBA) confirmation is the diagnostic approach used in most reports on seroprevalence. Second-generation tests are now available for use in the clinical setting; these assays detect additional antibodies, which increases their sensitivity.

Although seroprevalence for HCV is lower in the general population than for HBV, infection carries a much greater risk (as high as 50%) for the development of chronic hepatitis.[58] As with HBV, these patients are at increased risk for cirrhosis and hepatocellular carcinoma. Therefore, prevention of infection is essential. As with HBV, seroprevalence varies greatly depending on risk factors within the population tested. Prevalence in New York City blood donors from 1985–86, based on radioimmunoassay, was 0.9%;[54] when confirmed by RIBA, the positivity dropped to 0.14%.[27] A 1991 study of blood donor positivity in four US cities, using second-generation tests, found a range from 0.3 to 0.7%, which also decreased with RIBA confirmation.[28] However, the patient populations served by EMS personnel, particularly in urban areas, pose a much higher risk. A study of patients in a Baltimore emergency department found a seroprevalence of 18% in 2,523 subjects.[25]

HCV seroprevalence in health care workers tends to be elevated compared to blood donors, but not to the extent seen with HBV. Panlilio et al. reported an HCV seroprevalence of 0.9% in 770 surgeons; in contrast, 17% of the group had evidence of current or past HBV infection.[47] Health care workers in San Francisco were found to have a seroprevalence of 1.4%, which was approximately three times higher than the rate in the area's blood donors but much lower than their HBV seropositivity of 21.7%.[18]

Forseter et al. assessed antibody status in 51 dialysis staff members and found only 1 positive; that employee had a history of abnormal liver function tests that preceded her work in hemodialysis.[17] This low risk occurred despite contact with a patient population whose seroprevalence was 19% (24/125). Two studies have reported substantially higher seropositivities. A study of New York City dentists found an overall prevalence of 1.75% using first-generation tests; when examined by subgroups, however, 0.97% of 413 dentists were positive, but 9.3% of 43 oral surgeons had antibodies.[27] A study in a Japanese dialysis center found a seroprevalence of 17% in 129 health care workers and 53% in the patients.[39]

Two studies provide data on risk of contracting HCV after percutaneous exposure. Kiyosawa et al. used first-generation assays to study 110 employees who sustained needlestick injuries from an anti-HCV-positive source.[26] Four developed hepatitis; 3 of the 4 developed antibodies to HCV. In contrast, Mitsui et al. found a post-needlestick conversion rate of 10% in 68 health care workers exposed to blood containing HCV RNA.[39] The reason for the difference is thought to be related to the use of ELISAs, which assessed antibody response to four antigens combined with RNA confirmation in the latter study.

Research to date indicates that HCV is less infectious than HBV; this may reflect the lower viral load in the bloodstream.[26] However, there is clearly a risk of bloodborne transmission, and the increased potential for chronicity makes HCV prevention an important occupational health goal for the fire service. No vaccine is currently available, so protection is dependent on methods to decrease blood and body fluid exposure.

Hepatitis D

Hepatitis D virus, also known as the delta agent, is an incomplete virus that infects humans only in the presence of HBV. Infection can occur simultaneously with HBV or later as a superinfection in chronic carriers. It is similar to HBV in its clinical presentation, but seroprevalence is lower due to the requirement for HBV.[31] Hepatitis D infection in health care workers is not common, although it has been reported following needlestick exposure[31] and is probably under-reported when coinfection occurs.[32]

Human Immunodeficiency Virus

The human immunodeficiency virus (HIV) has been the driving force behind guidelines and regulations to prevent the transmission of bloodborne pathogens because, despite improved medical therapy, it remains ultimately fatal. As with all bloodborne pathogens, the seroprevalence of the population varies, based on extent of high-risk behavior. A recent study found a range of HIV seropositivity from 4.1% to 8.3% in patients to an inner-city emergency department who were transported by ambulance,[34] which confirms that, as with other bloodborne pathogens, some EMS personnel serve a high-risk population. Due to the sensitivity of the issue, less is known about HIV prevalence in health care workers than for HBV, although the available information indicates that HIV seropositivity is uncommon. Short and Bell cite studies that, when combined, found only 3 of 4,190 surgeons (0.07%) to be HIV-positive.[54] A study by Chamberland et al. also found an HIV prevalence of 0.07% in health care workers who donated blood at several sites across the United States.[13]

Occupational transmission occurs primarily from needlestick injuries. Studies have consistently shown that the risk of infection following percutaneous exposure to HIV is very small, approximately 0.3-0.4%.[21,55] The risk of transmission by mucous membrane contact is even less; 1 conversion has been reported in 1,107 prospectively studied health care workers (0.09%).[53] This decreased infectivity (compared to HBV) is thought to be due to lower concentrations of the virus in the blood of HIV-infected individuals.[4]

However, exposure in high-prevalence areas can result in a substantial working lifetime risk for HIV transmission. Marcus et al. studied EMS personnel in three U.S. cities with a high prevalence of acquired immunodeficiency syndrome (AIDS).[34] They interviewed subjects after shifts to ascertain the number of blood contacts and tested ambulance-transported patients for HIV in three receiving hospitals. They estimated an annual frequency of 0.2 percutaneous exposures per EMS worker, which, when multiplied by the 6.7% mean HIV seropositivity found in emergency patients, would result in 0.013 percutaneous HIV exposures. When multiplied by a transmission rate of 0.3% per percutaneous contact, an annual risk of 0.0039% for occupational HIV conversion is obtained; the risk over a 45-year working lifetime is 0.18%, or 180 per 100,000 workers. This is substantially higher than any of the similarly calculated risks from occupational carcinogens, which range from 1 to 100 per 100,000 workers.[51] However, the risk is much less in low-prevalence areas, as shown in a Marcus et al. study of emergency department employees in which they calculated a 0.002% annual frequency of percutaneous HIV exposure in an area with a mean HIV seropositivity of 0.47%.[35] This would result in a 45-year working lifetime risk of 27 per 100,000 workers.

Prevention

The CDC has published several guidelines that contain recommendations designed to reduce bloodborne pathogen transmission in the occupational setting.[4–6,8] OSHA incorporated these recommendations into its Occupational Exposure to Bloodborne Pathogens Standard.[45] These CDC and OSHA documents constitute a public health approach to the prevention of infection by bloodborne pathogens.

In addition to being essential resource materials for those involved in fire service infection control programs, the standard and CDC recommendations referred to therein are legal requirements for approximately 50% of all paid fire departments. As public employees, fire service personnel in the state, county, and municipal departments are not covered by federal OSHA laws. However, firefighters and other public employees are afforded protection in states that have their own state occupational safety and health plans. Employees in these states must be covered by standards at least as strict as those of federal OSHA. Federal firefighters are protected under federal OSHA.[16] In addition, a few states have voluntarily covered their public employees under a safety and health program for the OSHA hazard-specific standards which includes the Bloodborne Pathogens Standard.

The standard mandates the use of several mechanisms to protect workers and clearly defines the role of universal precautions and the hierarchy of controls in the management of exposure to blood and body fluids.[45] It calls for exposure control plans; sets requirements for limiting worker exposure through a combination of engineering controls, personal protective equipment, and worker training; and mandates that the hepatitis vaccination be offered to all at-risk employees. Finally, there is a provision for post-exposure evaluation so that workers who are exposed on the job can receive proper assessment of their risk and appropriate treatment.

Exposure Control. Initially, occupational infectious disease hazards for firefighters should be identified through exposure control plans that are developed by determining the exposure potential for different job classifications and activities within the fire department. Methods for limiting exposure to workers include a combination of engineering controls, work practices, and personal protective equipment. In the uncontrolled field setting characteristic of the fire service, engineering controls are difficult to employ. However, equipment designed to reduce the risk of needlesticks, such as resheathing needles, can be utilized and are increasingly available. Important work practices include universal precautions whenever exposure to bodily fluids is possible, hand washing after glove removal, avoidance of needle recapping, and proper needle and other waste disposal.

The use of personal protective equipment, such as gloves, gowns, face shields, and ventilation devices, is another essential means of exposure control for firefighters. Structural firefighting gloves are important in environments where contact with glass and sharp metal is possible, such as extrication of victims from automobiles. The 1999 Standard on Protective Clothing for Emergency Medical Operations, by the National Fire Protection Agency, is an industry standard on personal protective equipment for bloodborne pathogens and provides a resource for fire departments in the selection of such items.[42] It specifies minimum documentation, design criteria, performance criteria, and laboratory test methods for emergency medical protective clothing. The standard employs several rigorous tests to assess the effectiveness of clothing in preventing exposure to bacteria and viruses, while maintaining dexterity and integrity in an uncontrolled environment.

Employee Training. Employee training is critical in achieving compliance with the prevention methods outlined in the preceding section. In a 2-year study of firefighter/EMT personnel, the number of reported bloodborne pathogen exposures decreased by 51% in the second year of the study, which was attributed to employee education.[49] Training should include a discussion of the various bloodborne pathogens, including risks of transmission and preventive measures, as well as instruction in proper work practices. Education also provides an opportunity to reduce concerns regarding HIV infectivity by assuring workers that there is no evidence of transmission through casual contact with an infected individual or with saliva, tears, or urine (unless they are contaminated by blood). Reassurance that transmission during cardiopulmonary resuscitation is unlikely may also be beneficial.[14,15] The National Institute for Occupational Safety and Health (NIOSH) has published a curriculum guide on bloodborne pathogens for public-safety and emergency-response workers, which is a useful resource for employee training in the fire service.[43]

HBV Vaccination. Vaccination for HBV is required by OSHA for workers with occupational exposure to blood and other potentially infectious materials. All firefighters with emergency rescue responsibilities, either in medical response or during fire suppression, have the potential for occupational exposure and should be vaccinated. Synthetic vaccines are currently available and should be administered by intramuscular injection into the deltoid muscle initially and again at 1 and 6 months.[8] Obese individuals should be vaccinated with a 1.5-inch needle to ensure an intramuscular injection.[37] Routine booster doses are not currently recommended by the CDC.

Prevaccine testing is not likely to be cost-effective in firefighters. However, based on recent studies which assessed HBV response in the occupational setting, anti-HBs titers should be obtained after completion of the vaccine series, particularly in employees at high risk of nonresponse. Roome et al. found that 11.9% of 528 EMS personnel retested within 6 months of completion of the vaccine series had inadequate anti-HBs levels (<10 mIU/ml).[50] However, certain groups had much higher nonresponse rates. Age was an important determinant: 16.4% of 40- to 49-year olds were nonresponders, which increased to 22.1% for those from 50 to 59 years of age and 42.1% for persons over age 60.[50] Smoking also increased the nonresponse rate, as did a body mass index greater than 35. In addition, those tested between 3 and 6 months after completion of the vaccine series had a higher nonresponse rate, suggesting that the optimal time to assess anti-HBs status is within the first 3 months. Wood et al. found similar risk factors in health care workers.[59] The CDC recommends that nonresponders receive one to three repeat injections depending on response; as many as 50% will develop antibodies after three additional doses.[8]

Knowledge of initial antibody response in firefighters is helpful when determining treatment following an exposure. Some 25-60% of those who initially responded lose antibodies within 6-10 years but still are protected from clinical infection.[37] Therefore, it is important to know if a decreased antibody titer obtained in a postexposure evaluation represents an initial responder, whose titer has decreased but who is still likely to be afforded some protection, or an initial nonresponder who would need more specific treatment.

Evidence supporting the importance of vaccination is provided by HBV infection surveillance. Data in four sentinel counties throughout the United States showed a 75% decrease in cases of acute HBV infection in health care workers during the 7-year period from 1982 to 1988.[1] This decrease is thought to be due to vaccination and universal precautions. Lanphear et al. studied the rate of occupationally acquired clinical HBV infection at the University of Cincinnati Hospital from 1980 to 1989 and found a dramatic decrease in cases from 82 to 0 per 100,000 HCWs.[30] Employees with anti-HBs increased from 14% to 55% during this time (primarily due to vaccination); no change in source HBsAg-positivity was seen.

Postexposure Evaluation. It is essential that any employee who has an exposure incident (defined in the standard as a "specific eye, mouth, other mucous membrane, nonintact skin, or parenteral contact with blood or other potentially infectious materials"[45]) be evaluated as soon as possible. The purpose of this evaluation is to document the type of exposure, the health and HBV vaccination status of the employee, and, if possible, whether the source patient is infected with a bloodborne pathogen. Depending on the results, the employee can be given counseling regarding the various risks and treatments. Serologic evaluation of the source patient includes HBV and HIV (with the patient's consent). HCV testing may be performed depending on test availability, prevalence in the patient population, and specific source risk factors. Other suspected pathogens are assessed on an individual basis. Blood should be obtained on the employee to establish baseline serologic status.

To facilitate prompt evaluation, notification of emergency responders who were occupationally exposed to infectious diseases is now legally mandated in Subtitle B of the Ryan White Comprehensive AIDS Resources Emergency Act of 1990.[12] The act applies to all firefighters, both professional and volunteer. Infectious diseases that are potentially life-threatening, even in a healthy, susceptible host, and transmissible from person to person are included: infectious pulmonary tuberculosis, hepatitis B, HIV infection, diphtheria, meningococcal disease, hemorrhagic fevers, plague, and rabies. This law does not require a medical facility to test source patients for specific infectious diseases but does mandate a medical record review for clinical signs or laboratory evidence of disease.

The CDC protocol for HBV postexposure evaluation is shown in Table 1.[8] These guidelines, last published by the CDC in November 1991, are still applicable. Treatment is dependent on the hepatitis risk posed by the source and the vaccine status of the exposed employee. As a result, treatment recommendations vary substantially. For example, a known vaccine responder who is exposed to an HBsAg-positive source should be tested for current anti-HBs status; if adequate, no treatment is needed; if not, a booster vaccine dose is recommended. In contrast, exposed employees who are unvaccinated or known nonresponders should receive hepatitis B immune globulin (HBIG) as well as the vaccine. Follow-up of the employee's serologic status is also indicated. The OSHA Standard mandates treatment based on CDC recommendations and health care providers responsible for postexposure evaluation should consult the most recent CDC guidelines for complete details.

Postexposure evaluation for HCV involves obtaining baseline liver function tests and HCV serology with repeat testing within 6-9 months.[19] Postexposure prophylaxis

TABLE 1. Recommendations for Hepatitis B Prophylaxis Following Percutaneous Exposure

Exposed person	Treatment when source is found to be		
	HBsAg positive	HBsAg negative	Unknown or not tested
Unvaccinated	Administer HBIG × 1* and initiate hepatitis B vaccine	Initiate hepatitis B vaccine	Initiate hepatitis B vaccine
Previously vaccinated			
Known responder	Test exposed person for anti-HBs 1. If adequate, no treatment 2. If inadequate, hepatitis B vaccine booster dose	No treatment	No treatment
Known non-responder	HBIG × 2 or HBIG × 1, plus 1 dose of hepatitis B vaccine	No treatment	If known high-risk source, may treat as if source were HBsAg positive
Response unknown	Test exposed person for anti-HB§ 1. if inadequate HBIG × 1, plus hepatitis B vaccine booster dose 2. If adequate, no treatment	No treatment	Text exposed person for anti-HBs† 1. If inadequate, hepatitis B vaccine booster dose 2. If adequate, no treatment

*Hepatitis B immune globulin (HBIG) dose 0.06 ml/kg intramuscularly.
†Adequate anti-HBs is ≥10 milli-international units.
From the Centers for Disease Control: Hepatitis B virus: A comprehensive strategy for eliminating transmission in the United States through universal childhood vaccination. Recommendations of the Immunization Practices Advisory Committee (ACIP). MMWR 40(RR-13):1–25, 1991.

with immune serum globulin has been suggested but is not likely to be beneficial.[19] Interferon has also been suggested in this setting, but its efficacy is unknown and side effects may occur. Therefore, at present, there is no established treatment.[19,52] Evaluation after exposure to hepatitis D virus involves the treatment outlined for HBV, since the source patient carries both and prevention of HBV eliminates the possibility of hepatitis D virus infection.

After exposure to an HIV-positive source, the employee should have baseline serology obtained and be retested periodically for 6 months (e.g., at 6 weeks, 3 months). All acute illnesses, especially fever, rash, myalgia, fatigue, or lymphadenopathy, should be reported. During the follow-up period, especially the first 6-12 weeks, CDC recommendations for the prevention of HIV transmission should be adhered to, including abstinence or protected sex and avoidance of blood and organ donation.[5]

Information on the utility of zidovudine in post-HIV exposure chemoprophylaxis is still too limited to recommend it routinely. Since 1990, approximately 40% of health care workers in the CDC Cooperative Needlestick Surveillance study have elected to take zidovudine.[55] Seventy-five percent reported symptoms while on the medication; nausea, fatigue, and headache were most common. Thirty-one percent stopped the drug or lowered the dose as a result. More serious side effects in health care workers undergoing prophylaxis, including anemia, reversible peripheral neuropathy, and transient clinical hepatitis, are much less common but have been reported.[5] At least eight cases of zidovudine prophylaxis failure following needlestick exposure to HIV have occurred; the extent to which zidovudine resistance is involved in such cases is not known.[55] No benefit from zidovudine prophylaxis is evident in those studied thus far, although the number of enrolled workers needed to give the study the power to detect an effect is approximately twice the number enrolled in the first 10 years of this CDC study.[55] Animal work with the simian immunodeficiency virus in primates has not established postexposure efficacy either.[19]

However, a CDC case-control study of HIV-exposed health care workers found an increased risk for viral transmission in health care workers who did not take zidovudine.[19] In addition, there are currently no other prophylaxis alternatives for HIV infection. Therefore, the decision to treat must be an individual one based on issues such as the type of exposure and the time since exposure. The employee should be counseled on the risks and benefits and allowed to make an informed decision. If prophylaxis is chosen, hematologic and hepatic monitoring should be performed while on zidovudine.

Lastly, employees should be aware of the need to consult the fire department's occupational health professional if they have health conditions, such as open skin lesions or immunologic abnormalities, that may place them at greater risk from workplace exposures. Accommodations may be necessary in certain circumstances. Care should be taken not to discriminate against such employees.

AIRBORNE PATHOGENS

Tuberculosis

After several decades of declining incidence, TB has reemerged as a serious public health threat. From 1985 to 1992, a 20.1% increase in the number of reported TB cases occurred, primarily in urban areas.[10] The incidence is especially high in AIDS patients, the homeless, prison inmates, recent immigrants from countries where TB is common, and parenteral drug users. In addition, multidrug-resistant TB has been reported in several hospitals and correctional facilities. This resistant TB is an even greater public health threat because treatment is difficult, especially in immunocompromised individuals.

TB is a particularly hazardous occupational exposure because transmission is via the airborne route, which is difficult to control in health care facilities; control in emergency response settings is even more problematic. Infectious individuals transmit the mycobacterium through airborne particles or droplet nuclei, produced by coughing, sneezing, or talking. In the absence of adequate ventilation, these respirable droplets can survive suspended in the air for several hours, resulting in continued opportunity for exposure.

Inhalation of infectious airborne particles more commonly results in latent infection than active disease. The purified protein derivative (PPD) test generally becomes positive within 12 weeks in these individuals, although these individuals are not infectious. However, they have a 10% chance of subsequently developing active TB; this risk is highest in the 2 years following exposure.[11] The risk is much higher in individuals with HIV infection and similar immune system abnormalities.

Furthermore, contact does not have to be prolonged, as previously thought, for infection to occur. This finding has important implications for EMS personnel, who may have limited but intense exposure to infectious patients in ambulances. Markowitz, in a recent summary of TB epidemiology in health care workers, cites outbreak investigations that found PPD test conversion rates of 40-80% in exposed employees after only a few hours of intense contact with an infectious patient.[38] In one outbreak, both PPD conversions and active TB were reported in personnel from exposure to an infectious patient who was intubated while in the emergency department but spent a total of only 4 hours there.[20] A recent summary of annual PPD conversion rates in workers from both health care and correctional facilities revealed wide variation, from rates of 1-3%, which are similar to those in the community, to 11% in pulmonary fellows.[2] Postexposure conversion rates as high as 77% have been reported in staff exposed during the bronchoscopy of a patient with active disease.[2] CDC investigations of multidrug-resistant outbreaks have documented at least 100 PPD conversions in health care workers, 17 active multidrug-resistant TB cases, and 7 deaths.[33]

EMS personnel are often involved in treating and transporting individuals in high-risk groups and thus are at increased risk for occupationally acquired TB. These workers may be exposed in small spaces with inadequate ventilation (e.g., the back of ambulances), which serves to increase particle concentration. Information on annual PPD conversion rates in EMS personnel is limited; however, at least one fire department has instituted a PPD screening program that should yield important information on the extent of occupational risk in this workforce (Pat Bielecki, Phoenix Fire Department, personal communication).

Prevention. The most effective means of controlling TB are early detection, isolation, and treatment. The CDC has been instrumental in providing prevention guidelines. Interagency collaboration among the CDC, OSHA, and NIOSH has resulted in information specifically for the protection of health care workers. The most recent CDC guidelines, published in 1994, emphasize several essential components in a prevention strategy.[11] For EMS personnel, the first component is a written infection control plan that assesses risk for the specific fire department and outlines worker-protection strategies. The plan should contain protocols for identifying patients who may have active TB.

The second component consists of exposure control measures, of which four types are applicable for firefighter/EMS personnel. Engineering controls to increase general ventilation include operating the heating and air-conditioning systems on a nonrecirculating cycle and opening ambulance windows when feasible. Administrative controls include limiting the number of persons exposed to suspected TB patients. Work practice controls, such as placing a surgical mask on the patient, are also recommended. Finally, and most importantly for emergency responders, personal protective equipment in the form of a respirator should be used whenever exposure to a patient with suspected TB occurs. The CDC guidelines specifically mention the need for respiratory protection in emergency transport vehicles.

Respirators used for the prevention of TB should meet four performance criteria:[11]
- Filter at least 95% of particles 1 μm in size
- Have the ability to be fit-tested to ensure face-seal leakage of 10% or less
- Have at least three sizes available to accommodate different face sizes
- Allow verification of facepiece fit with each use

Supplement 4 of the guidelines should be consulted for additional information.[11] Currently, only respirators with high-efficiency particulate air (HEPA) filters are NIOSH certified to meet these performance criteria.[23] However, NIOSH is revising its certification process, which is expected to result in the manufacture of new respirators that will be similar to current dust-mist-fume types in terms of cost and convenience but as efficient as HEPA in particle filtering.[23,44]

Worker education is the next essential component in a TB-prevention program. Basic TB information and general principles of infection control as well as specifics for each workplace should be included. Workers must be trained to identify patients who may have active TB so that proper protective measures can be instituted. The importance of PPD screening programs should be emphasized. Training may also be an opportunity to make employees aware that immunocompromised individuals are at increased risk for TB and should seek guidance from the occupational health provider or their physician regarding the need for job accommodations.

A screening program based on PPD skin testing should be instituted for all firefighter/EMS personnel with patient contact. A two-step test is recommended in the preplacement evaluation of workers who have not had a documented negative test in the preceding 12 months to distinguish boosting from recent conversion.[11] The PPD test should be repeated every 6 to 12 months depending on frequency of exposure. Detailed, updated guidance for test interpretation is provided in the 1994 CDC guidelines (Table 2).[11] In addition to size of PPD induration, the action to be taken depends on the health status and exposure history of the worker. Interpretation in health care workers may also vary depending on TB prevalence in the patient population served. Results of the PPD program should be analyzed regularly as an indication of the success of the exposure control plan. Further investigation is necessary if elevated conversion rates are found or if occupationally acquired TB occurs.

Employees whose PPD tests are determined to be positive should be evaluated clinically and have a chest radiograph obtained. Workers should also be evaluated if unprotected exposure occurs or if symptoms of active disease (cough ≥ 3 weeks, weight loss, night sweats, fever, bloody sputum) develop. If active TB is diagnosed, the affected firefighter needs to be placed off work until he or she is no longer infectious (this does not apply to those on preventive therapy for PPD conversion). Therapy for active disease and prevention are dependent on prevalence of multidrug-resistant TB in the patient population served. The CDC guidelines contain a detailed discussion of the many factors that enter into interpretation and treatment decisions and should be consulted for individual case management.[11]

In October 1993, OSHA issued an enforcement policy and procedures for occupational exposure to TB.[46] As with the Bloodborne Pathogens Standard, public employees are covered in states with their own OSHA state plans. The document is based on previous CDC recommendations and will ultimately be replaced by an OSHA compliance directive based on the 1994 CDC guidelines discussed earlier. If employers are found to be noncompliant, they may be cited under the OSHA General Duty Clause, Section 5(a)(1) of the OSH Act, which states that employers must provide a place of employment that is free from recognized hazards likely to cause death or serious physical harm. In addition to the CDC elements, OSHA specifies that workers' compensation and disability shall be provided to any employee with occupationally related TB and that positive PPD tests (other than preplacement) be reported on the OSHA 200 log. The Ryan White Act provides for automatic notification of the emergency response employee if the transported patient is found to have infectious pulmonary tuberculosis.

Influenza

The influenza viruses are highly infectious, and substantial segments of the population are susceptible to repeated infections due to frequent antigenic variation in these viruses. Elderly persons and those with underlying chronic medical problems are at increased risk for serious complications, such as pneumonia. Influenza is a concern for health care professionals because of increased exposure through patient contact and for risk of transmission to patients, particularly those in high-risk categories. The CDC recommends vaccination of health care workers who have close contact with high-risk patients and vaccine consideration for workers who provide essential public functions.[9] An additional risk for firefighters is the close living conditions in the fire station, which makes avoidance of an influenza outbreak important. Yearly vaccination of all firefighters is advisable.

TABLE 2.   Summary of Interpretation of Purified Protein Derivative-Tuberculin Skin-Test Results

1. An induration of ≥5 mm is classified as positive in:
    - persons who have human immunodeficiency virus (HIV) infection or risk factors for HIV infection but unknown HIV status;
    - persons who have had recent close contact* with persons who have active tuberculosis;
    - persons who have fibrotic chest radiographs (consistent with healed TB).

2. An induration of ≥10 mm is classified as positive in all persons who do not meet any of the criteria above but who have other risk factors for TB, including:

    High-risk groups—
    - injecting-drug users known to be HIV seronegative;
    - persons who have other medical conditions that reportedly increase the risk for progressing from latent TB infection to active TB (e.g., silicosis; gastrectomy or jejuno-ileal bypass; being ≥10% below ideal body weight; chronic renal failure with renal dialysis; diabetes mellitus; high-dose corticosteroid or other immuno-suppressive therapy; some hematologic disorders, including malignancies such as leukemias and lymphomas; and other malignancies);
    - children <4 years of age.

    High-prevalence groups—
    - persons born in countries in Asia, Africa, the Caribbean, and Latin America that have high prevalence of TB;
    - persons from medically underserved, low-income populations;
    - residents of long-term-care facilities (e.g., correctional institutions and nursing homes);
    - persons from high-risk populations in their communities, as determined by local public health authorities.

3. An induration of ≥15 mm is classified as positive in persons who do not meet any of the above criteria.

4. Recent converters are defined on the basis of both size of induration and age of the person being tested:
    - ≥10 mm increase within a 2-year period is classified as a recent conversion for persons <35 years of age;
    - ≥15 mm increase within a 2-year period is classified as a recent conversion for persons ≥35 years of age.

5. PPD skin-test results in health care workers (HCWs)
    - In general, the recommendations in sections 1, 2, and 3 of this table should be followed when interpreting skin-test results in HCWs.
    However, the prevalence of TB in the facility should be considered when choosing the appropriate cut-point for defining a positive PPD reaction. In facilities where there is essentially no risk for exposure to *Mycobacterium tuberculosis* (i.e., minimal- or very low-risk facilities [Section II.B]), an induration ≥15 mm may be a suitable cut-point for HCWs who have no other risk factors. In facilities where TB patients receive care, the cut-point for HCWs with no other risk factors may be ≥10 mm.
    - A recent conversion in an HCW should be defined generally as a ≥10 mm increase in size of induration within a 2-year period. For HCWs who work in facilities where exposure to TB is very unlikely (e.g., minimal-risk facilities), an increase of ≥15 mm within a 2-year period may be more appropriate for defining a recent conversion because of the lower positive-predictive value of the test in such groups.

---

*Recent close contact implies either household or social contact or unprotected occupational exposure similar in intensity and duration to household contact.
From the Centers for Disease Control and Prevention: Guidelines for preventing the transmission of *Mycobacterium tuberculosis* in health-care facilities. MMWR 43(No. RR-13):1–133, 1994.

OTHER PATHOGENS

Measles, mumps, and rubella are viral pathogens transmitted by large respiratory droplets and thus generally require close contact with an infected source for transmission to occur. The incidence of measles has recently increased sharply from previous postvaccine levels. In 1990, 22% of CDC-reported cases occurred in adults ≥ 20 years of age, one-quarter of whom required hospitalization.[7] Rubella is a concern for pregnant women due to its teratogenic potential. These illnesses pose occupational risks to firefighters in a variety of ways. Exposure may result in lost work time until presence or absence of infection is determined, illness may occur, and close living quarters in the station house may result in rapid spread of disease.

The Immunization Practices Advisory Committee of the CDC recommends vaccination of nonimmune health care workers who have potential for exposure to infected patients or, in the case of rubella, pregnant women.[7] For the reasons discussed earlier, it is advisable to include firefighters in these vaccine recommendations. If not already immune (previous vaccination in accordance with current CDC recommendations, physician documented illness, or current serologic evidence of immunity), firefighters should be vaccinated, preferably when first hired. Measles, mumps, and rubella (MMR) vaccine can be given to those needing immunity to all three. Individuals born before 1957 are generally thought to have natural immunity to measles and mumps. However, because a third of health care workers who developed measles between 1985 and 1989 were born before 1957, providing at least one dose of vaccine to all employees who do not have proof of immunity may be beneficial.[7]

Meningococcal meningitis is transmitted via respiratory droplets, which also involves close patient contact. Although it is not a common infection, it is listed in the Ryan White Act because it can be lethal in healthy individuals, especially if prompt treatment is not instituted. Symptoms and signs of meningitis include acute onset of fever, headache, stiff neck, and vomiting. A 2-day course of rifampin has been the traditional choice for chemoprophylaxis following exposure, although ceftriaxone and ciprofloxacin are newer options that are also effective and increasingly used.

Hepatitis A virus (HAV) is transmitted via the fecal-oral route and causes acute hepatitis but does not establish a chronic carrier state. Most studies have not found an increased risk for HAV infection in health care workers,[32] and it is not a significant occupational hazard for firefighters or EMS personnel. However, a case did occur in a firefighter who was exposed to infected fecal materials while cleaning contaminated tools after a garbage chute fire (Deborah Owen, MD, MPH, San Francisco Fire Department, personal communication). Should exposure to HAV-contaminated feces occur, administration of immune globulin is recommended.[6] Prevention includes standard infection control procedures, such as glove use and strict handwashing when handling fecal material. A vaccine is now available, but its use in adults is advocated primarily for travelers to endemic areas.[36]

Exposure to many other infectious agents may occur in firefighters. **Lyme disease** is a potential risk for wildland firefighters who may spend days in tick-infested wilderness areas. Preventive measures include the use of protective clothing and tick repellent. Employee education and postexposure medical evaluations are also important. The **herpes simplex virus** is spread through direct contact with infected secretions. Preventive measures include personnel protective equipment, such as gloves for direct contact with potentially infectious lesions, and ventilation devices when performing cardiopulmonary resuscitation (CPR). Possible transmission of *Streptococcus pyogenes* *(toxic strep)* from a patient to a firefighter during CPR has also been reported.[57]

Compliance with general infection control measures reduces the risk of these occupational illnesses, and workers should receive annual training to reinforce these concepts. Specific therapies may be indicated on an individual basis, and a health care professional with expertise in occupational infectious disease should be available for consultation as well as input into the infection control program

ADDITIONAL RESOURCES

Professional standards, such as the National Fire Protection Association (NFPA) documents, are applied to fire departments through voluntary adoption by the municipality. NFPA 1500, Standard on Fire Department Occupational Safety and Health Programs, was developed as a consensus standard for the fire service.[41] One of the standard's requirements is that fire departments limit or prevent the exposure of members to occupational infectious diseases and provide treatment and vaccinations when appropriate. The NFPA 1500 Standard recommends consultation with medical professionals on measures to limit the exposure of firefighters to contagious diseases.

NFPA 1581, Standard on Fire Department Infection Control Programs, addresses the provision of minimum requirements for infection control practices within a fire department.[40] An infection control program designed for the fire service is outlined. Initial program components include a policy statement, training program, designated infection control personnel, and exposure incident reports. Safe work practices, proper use of personal protective equipment, and the decontamination and disposal of contaminated waste and equipment are also addressed. Medical issues including administration of the hepatitis vaccine to all personnel, maintenance of a confidential health file, and prompt provision of medical evaluation following an exposure are required. Finally, consideration of infection control measures in the design of fire station bathrooms, kitchens, sleeping areas, and laundry facilities are included.

Individuals and health care professionals responsible for infection control program elements should also be aware of the United States Fire Administration as a resource. Lastly, local union presidents may request information on infectious disease issues or the most recent CDC, OSHA, and NIOSH publications from the Occupational Health and Safety Department of the International Association of Firefighters.

REFERENCES

1. Alter MJ, Hadler SC, Margolis HS, et al: The changing epidemiology of hepatitis B in the United States. JAMA 263:1218–1222, 1990.
2. Bowden KM, McDiarmid MA: Occupationally acquired tuberculosis: What's known. J Occup Med 36:320–325, 1994.
3. Brunacini AV: 1993 Phoenix Fire Department National Survey on Fire Department Operations in the United States and Canada. Phoenix, AZ, Phoenix Fire Department, 1993.
4. Centers for Disease Control: Guidelines for prevention of transmission of human immunodeficiency virus and hepatitis B virus to health-care and public-safety workers. MMWR 38(S-6):1–37, 1989.
5. Centers for Disease Control: Public health service statement on management of occupational exposure to human immunodeficiency virus, including considerations regarding zidovudine postexposure use. MMWR39(R-1):1–14, 1990.
6. Centers for Disease Control: Protection against viral hepatitis: Recommendations of the Immunization Practices Advisory Committee (ACIP). MMWR 39(RR-2):1–26, 1990.
7. Centers for Disease Control. Update on adult immunization: Recommendations of the Immunization Practices Advisory Committee (ACIP). MMWR 40(RR-12): 1–7, 1991.
8. Centers for Disease Control: Hepatitis B virus: A comprehensive strategy for eliminating transmission in the United States through universal childhood vaccination. Recommendations of the Immunization Practices Advisory Committee (ACIP). MMWR 40(RR-13):1–25, 1991.
9. Centers for Disease Control: Prevention and control of influenza: Recommendations of the Immunization Practices Advisory Committee (ACIP). MMWR 41(RR-9), 1992.
10. Centers for Disease Control and Prevention: Tuberculosis morbidity—United States, 1992. MMWR 36:696–704, 1993.

11. Centers for Disease Control and Prevention: Guidelines for preventing the transmission of *Mycobacterium tuberculosis* in health-care facilities. MMWR 43(RR-13):1–133, 1994.
12. Centers for Disease Control and Prevention: Implementation of Provisions of the Ryan White Comprehensive AIDS Resources Emergency Act Regarding Emergency Response Employees: Public Law 101-381, Sec. 411. Fed Reg 59:13418–13428, 1994.
13. Chamberland ME, Petersen LR, Munn VP, et al: Human immunodeficiency virus infection among health care workers who donate blood. Ann Intern Med 121:269–273, 1994.
14. Cummins RO: Infection control guidelines for CPR providers. JAMA 262:2732–2733, 1989.
15. Emergency Cardiac Care Committee, American Heart Association: Risk of infection during CPR training and rescue: Supplemental guidelines. JAMA 262:2714–2715, 1989.
16. Executive Order 12196: Occupational Safety and Health Programs for Federal Employees. Fed Reg 45:12769, 1980.
17. Forseter G, Wormser GP, Adler S, et al: Hepatitis C in the health care setting: II. Seroprevalence among hemodialysis staff and patients in suburban New York City. Am J Infect Control 21:5–8, 1993.
18. Gerberding JL: Incidence and prevalence of human immunodeficiency virus, hepatitis B virus, hepatitis C virus, and cytomegalovirus among health care personnel at risk for blood exposure: Final report from a longitudinal study. J Infect Dis 170:1410–1417, 1994.
19. Gerberding JL: Management of occupational exposures to blood-borne viruses. N Engl J Med 332:444–451, 1995.
20. Haley CE, McDonald RC, Rossi L, et al: Tuberculosis epidemic among hospital personnel. Infect Control Hosp Epidemiol 10:204–210, 1989.
21. Henderson DK, Fahey BJ, Willy M, et al: Risk for occupational transmission of human immunodeficiency virus type 1 (HIV-1) associated with clinical exposures. Ann Intern Med 113:740–746, 1990.
22. Janssen RS, St. Louis ME, Satten GA, et al: HIV infection among patients in U.S. acute care hospitals: Strategies for the counseling and testing of the hospital patients. N Engl J Med 327:445–452, 1992.
23. Jarvis WR, Bolyard EA, Bozzi CJ, et al: Respirators, recommendations, and regulations: The controversy surrounding protection of health care workers from tuberculosis. Ann Intern Med 122:142–146, 1995.
24. Jui J, Modesitt S, Fleming D, et al: Multicenter HIV and hepatitis B seroprevalence study. J Emerg Med 8:243–251, 1990.
25. Kelen GD, Green GB, Purcell RH, et al: Hepatitis B and hepatitis C in emergency department patients. N Engl J Med 326:1399–1404, 1992.
26. Kiyosawa K, Sodeyama T, Tanaka E, et al: Hepatitis C in hospital employees with needlestick injuries. Ann Intern Med 115:367–369, 1991.
27. Klein RS, Freeman K, Taylor P, Stevens CE: Occupational risk for hepatitis C virus infection among New York City dentists. Lancet 338:1539–1542, 1991.
28. Kleinman S, Alter H, Busch M, et al: Increased detection of hepatitis C virus (HCV)-infected blood donors by a multiple-antigen HCV enzyme immunoassay. Transfusion 32:805–813, 1992.
29. Kunches LM, Crave DE, Werner BG, Jacobs LM: Hepatitis B exposure in emergency medical personnel: Prevalence of serologic markers and need for immunization. Am J Med 75:269–272, 1983.
30. Lanphear BP, Linnemann CC, Cannon CG, DeRonde MM: Decline of clinical hepatitis B in workers at a general hospital: Relation to increasing vaccine-induced immunity. Clin Infect Dis 16:10–14, 1993.
31. Lettau LA, Alfred HJ, Glew RH, et al: Nosocomial transmission of delta hepatitis. Ann Intern Med 104:631–635, 1986.
32. Lettau LA: The A, B, C, D, and E of viral hepatitis: Spelling out the risks for healthcare workers. Infect Control Hosp Epidemiol 13:77–81, 1992.
33. Maloney SA, Pearson ML, Gordon MT, et al: Efficacy of control measures in preventing nosocomial transmission of multidrug-resistant tuberculosis to patients and health care workers. Ann Intern Med 122:90–95, 1995.
34. Marcus R, Srivastava PU, Bell DM, et al: Occupational blood contact among prehospital providers. Ann Emerg Med 25:776–779, 1995.
35. Marcus R, Culver DH, Bell DM, et al: Risk of human immunodeficiency virus infection among emergency department workers. Am J Med 94:363–370, 1993.
36. Margolis HS, Alter MJ: Will hepatitis A become a vaccine preventable disease? Ann Intern Med 122:464–465, 1995.
37. Margolis HS, Presson AC: Host factors related to pool immunogenicity of hepatitis B vaccine in adults: Another reason to immunize early. JAMA 270:2971–2972, 1993.
38. Markowitz SB: Epidemiology of tuberculosis among health care workers. Occup Med State Art Rev 9:589–608, 1994.
39. Mitsui T, Iwano K, Masuko K, et al: Hepatitis C virus infection in medical personnel after needlestick accident. Hepatology 16:1109–1114, 1992.
40. National Fire Protection Association: NFPA 1581 Standard on Fire Department Infection Control Program. Quincy, MA, NFPA, 1991.

41. National Fire Protection Association: NFPA 1500 Standard on Fire Department Occupational Safety and Health Program. Quincy, MA, NFPA, 1992.
42. National Fire Protection Association: NFPA 1999 Standard on Protective Clothing for Emergency Medical Operations. Quincy, MA, NFPA, 1992.
43. National Institute for Occupational Safety and Health: A Curriculum Guide for Public-Safety and Emergency-Response Workers: Prevention of Transmission of Human Immunodeficiency Virus and Hepatitis B Virus. Cincinnati, NIOSH, 1989, DHHS publication 89–108.
44. National Institute for Occupational Safety and Health: Respiratory protection devices; proposed rule: Health-care-facility workers potentially exposure to tuberculosis. Fed Reg 26850–26889, May 24, 1994.
45. Occupational Safety and Health Administration: Occupational exposure to bloodborne pathogens: Final Rule, 29 CFR Part 1910.1030. FR 56:64052, 1991.
46. Occupational Safety and Health Administration: Enforcement Policy and Procedures for Occupational Exposure to Tuberculosis. Washington, DC, OSHA, 1993.
47. Panlilio AL, Shapiro CN, Schable CA, et al: Serosurvey of human immunodeficiency virus, hepatitis B virus, and hepatitis C virus infection among hospital-based surgeons. J Am Coll Surg 180:16–24, 1995.
48. Pepe PE, Hollinger FB, Troisi CL, Heiberg D: Viral hepatitis risk in urban emergency medical services personnel. Ann Emerg Med 15:454–457, 1986.
49. Reed E, Daya MR, Jui J, et al: Occupational infectious disease exposures in EMS personnel. J Emerg Med 11:9–16, 1993.
50. Roome AJ, Walsh SJ, Cartter ML, Hadler JL: Hepatitis B vaccine responsiveness in Connecticut public safety personnel. JAMA 270:2931–2934, 1993.
51. Rosenberg J, Becker CE, Cone JE: How an occupational medicine physician views current bloodborne disease risks in health-care workers. Occup Med State Art Rev 4:3–6, 1989.
52. Schiff ER: Hepatitis C among health care providers: Risk factors and possible prophylaxis. Hepatology 16:1300–1301, 1992.
53. Short LJ, Bell DM: Risk of occupational infection with blood-borne pathogens in operating and delivery room settings. Am J Infect Control 21:343–350, 1993.
54. Stevens CE, Taylor PE, Pindyck J, et al: Epidemiology of hepatitis C virus: A preliminary study of volunteer blood donors. JAMA 263:49–53, 1990.
55. Tokars JI, Marcus R, Culver DH, et al: Surveillance of HIV infection and zidovudine use among health care workers after occupational exposure to HIV-infected blood. Ann Intern Med 118:913–919, 1993.
56. Valenzuela TD, Hook EW, Copass MK, Corey L: Occupational exposure to hepatitis B in paramedics. Arch Intern Med 145:1976–1977, 1985.
57. Valenzuela TD, Hooton TM, Kaplan EL, Schlievert P: Transmission of 'toxic strep' syndrome from an infected child to a firefighter during CPR. Ann Emerg Med 20:90–92, 1991.
58. Weiland O, Schvarcz R: Hepatitis C: Virology, epidemiology, clinical course, and treatment. Scand J Gastroenterol 27:337–342, 1992.
59. Wood RC, MacDonald KL, White KE, et al: Risk factors for lack of detectable antibody following hepatitis B vaccination of Minnesota health care workers. JAMA 270:2935–2939, 1993.

RICHARD GIST, PhD
S. JOSEPH WOODALL

OCCUPATIONAL STRESS IN CONTEMPORARY FIRE SERVICE

From Johnson County Community
 College
Overland Park, Kansas (RG)
 and
Eastern Arizona College
Payson, Arizona (SJW)

Reprint requests to:
Richard Gist, PhD
Director, Social Sciences and
 Social Services
Johnson County Community
 College
12345 College Blvd.
Overland Park, KS 66210-1299

Firefighting ranks among the most stressful occupations in the American workplace; this truism is so intuitively obvious and has been so often stated[28] that it seems almost gratuitous to repeat it here. Yet, despite the ubiquitous permeation of this concept throughout both professional and colloquial venues—or, indeed, perhaps because of that permeation—the quality of information and scholarship surrounding the matter has been subject to extreme variation, and significant limitations in both understanding of related phenomena and intervention therein have seemingly become the rule.

CONTEMPORARY VIEWS OF FIREFIGHTER STRESS

Recognition of stress as a significant factor affecting firefighter health and safety no longer meets "macho" resistance with the industry; no longer do stress management programs encounter denial so intense as to demand or justify missionary zeal to secure endorsement and adoption. Indeed, references to stress and stress management have gone in little more than a decade's time from an avant garde rarity in fire service trade literature and instruction to what seems a nearly gratuitous frequency. Most of this upsurge, however, has been focused on a relatively singular and narrow construction of occupational stress and occupational stressors (i.e., critical incident stress) and a specific popularized model of intervention (i.e., critical incident stress debriefing).

These treatments have recited a relatively uniform, if peculiarly labyrinthine, catechism regarding the nature, sources, and impact of these presumed occupational stressors and have been especially uniform in their reliance on a single author whose work has remained well outside the refereed literature of psychology and its related disciplines.[33-39] Much like the suicide prevention and crisis intervention movement of a quarter century ago,[11] this phenomenon has arisen primarily as a grass roots amalgam of a self-anointed laity, self-generated paraprofessionals, semiprofessionals from the more distal end of the education and practice continuum, and a few charismatic leaders–highly revered within the movement itself but largely unrecognized and unestablished outside its boundaries. It is only very recently, well after its growth into a substantial cottage industry with a disturbingly strong foothold in the fire and rescue professions, that solid academic scrutiny has been attracted.

Not surprisingly, this scrutiny has been neither entirely welcomed nor embraced within the "movement" itself. The various descriptors evolved by Janis[25] in his model of "groupthink" or Hoffer's[24] classic descriptors of "the true believer" seem oddly reflected in the sociology of information and its dissemination with respect to what would otherwise objectively seem to be simply a bellwether of paradigmatic shifts regarding the nature of the enterprise, the personnel who conduct it, and the changes in management philosophy and approaches such shifting compels. Accordingly, the first critical treatments to appear have met with substantial hesitance and even overt resistance and have had to wrest their way through extreme editorial scrutiny from the very trade journals that unquestioningly published the early self-veneration of the movement's founders and supplied ready complicity in the uncritical promotion of their aggrandized claims of unique and extraordinary merit.[20]

The core contentions on which this enterprise has been constructed are derived from the essential assumption that occupational events function on an individual level as psychic traumata, wounding the psyche of the individuals confronting them so as to disrupt the capacity of such persons to function normally in the aftermath.[36-39] The premise that these exposures, if not contravened through direct and focused interventions, will lead to posttraumatic stress disorder and related psychiatric maladies[35] has been so often repeated as to be presumed ubiquitous and has become central to arguments supporting ever more elaborate intervention schemata. A quick review of the actual literature base, however, raises significant questions about both this derivation and its construction.

POSTTRAUMATIC STRESS DISORDER

The diagnosis of posttraumatic stress disorder first appeared in the nomenclature of the American Psychiatric Association with the third revision if its *Diagnostic and Statistical Manual* (DSM-III) in 1980. Its derivation flowed in large part from early Freudian concepts of "psychic traumatization" through exposure to events that might cause overwhelming fear and consequent neurotic defense to the anxiety thereby provoked, and specifically from their application to "traumatic neurosis" in certain warfare combatants, particularly those presenting "shell-shock," "trigger-finger paralysis," or similar variations of more classical hysterical fugues, dissociations, or conversions. These conditions were ordinarily believed to be transitory and to resolve when impingement of the salient stressor was removed.

That specific construction had been carried into the second version of the *Diagnostic and Statistical Manual* (DSM-II), where such reactions were classified as transient situational disturbances. The defining characteristics of this category included "more or less transient disorders of any severity (including those of psychotic proportions) that occur in individuals without any apparent underlying mental disorders and that represent an acute reaction to overwhelming environmental stress."[2] These were to be differentiated from manifestations of underlying true neuroses exacerbated by context and from malingering or occupational maladjustment—a category reserved for "psychiatrically normal individuals who are grossly maladjusted in their work."[2] The differential diagnosis was based in good measure on the persistence of symptoms once exposure was terminated, to wit:

> If the patient has good adaptive capacity his symptoms usually recede as the stress diminishes. If, however, the symptoms persist after the stress is removed, the diagnosis of another mental disorder is indicated.[2]

The appearance of the specific category in the third revision (DSM-III) may represent in part the contributions of at least two salient features of the zeitgeist surrounding psychiatry at that juncture. The first element was reflected in a deliberate effort to eliminate etiologic and theoretical assumptive bases nested within the nomenclature—hence the disappearance of terms such as "neurosis" and the restructuring of the system along basically descriptive parameters. The second element was reflected in a rather phenomenal expansion in the range of behaviors considered classifiable and/or pathognomonic, as witnessed by the expansion of the DSM itself from a small, spiral-bound guidebook of 134 pages into a substantial volume approaching 2 inches in its paperbound form.

Posttraumatic stress disorder emerged in greatest measure to accommodate reactions reported within a certain subpopulation of Vietnam veterans. These reactions were characterized by intrusive images in the form of flashbacks or recurrent visualizations, hyperarousal and startle responses, and similar persistent features associated with recall of the experience. Diagnosis required onset only after exposure to a profound stressor "outside the range of ordinary human experience" and persistence of these symptoms for at least 6 months.[3] These criteria were admittedly arbitrary and in reality represented little more than the consensus of a subcommittee.

Between the publication of the third revision of the *Diagnostic and Statistical Manual* (DSM-III) and its subsequent further revision (DSM-III-R) in 1987, interest in this particular diagnosis grew substantially in at least two distinct arenas. Specialized treatment facilities for Vietnam veterans grew within the Veterans Administration health care system and beyond, establishing a foundation for those pursuing combat-related stress issues. Nearly simultaneously, other advocates and practitioners, especially those connected to specific social movements advocating rights and recognition for ostensibly victimized elements of society, sought application of the diagnosis and the label to persons exposed to other life-threatening "traumata" (e.g., rape, violent crime, abuse, natural disaster). This led in part to substantial expansion in the range of applicable stressors that might be said to precipitate such reactions and to deemphasis of both the persistence criterion and of differential diagnosis on the basis of preexistent morbidity. Accordingly, arguments have proliferated for application of this diagnosis to any perceived inhibition of adjustment to the impact of most any of life's more unfortunate experiences or circumstances.

Many of these arguments, though, may hinge as much on rhetoric as on any substantive empirical base. Indeed, study of more than 140,000 survivors of the Nazi holocaust who emigrated to the United States suggested not only that these individuals were not prone to manifest such social and individual maladjustments but that they instead presented an overall picture of social integration and individual resiliency that exceeded that of a nonexposed age cohort of American Jews.[23] Metaanalysis and review of studies of Vietnam veterans[27] suggested that the minority of combat-exposed personnel who ultimately experienced severe adjustment reactions might well have borne premorbid diagnoses not revealed in the retrospective venue of diagnosis and study that has characterized the evolution of this particular assignation. Even at that, however, it must be noted that either of these groups endured prolonged, profound, and personally threatening impingements—certainly, their exposure was not of a singular critical episode, nor was it by any means the simple exacerbation of ordinary challenges in working and living.

Despite continuing suggestions from clinical workers that case-exclusionary criteria should be weakened still further,[6,45] more rigorous and arguably more objective academic constructions of the underlying issues point in quite another direction. McFarlane's extensive work[31,32] on the etiology and genesis of posttraumatic morbidity suggested that nature or extent of exposure, while perhaps moderately predictive of perceived distress, was not a strong predictor of lasting psychiatric sequelae; the latter condition was predicted only by premorbid adjustment. Similarly, Cook and Bickman[7] reported that exposure determined only the intensity of the immediate distress but held no predictive effect respecting any longer-term impacts. Conversely, a strong line of research has directly questioned the presumed linkages between life stressors and illness in general[44] and the impact of major life events in particular;[51] the latter line, in keeping with a more interactive social and cognitive model of stress proposed by Lazarus and Folkman,[12] has demonstrated that major life events exert little if any disruptive influence directly but rather operate through exacerbation of daily hassles, holding no necessary connection (other than temporal association and propinquity) to the presumed major stressor.

The best synopsis, then, of the current construction of specific occupational events precipitating posttraumatic stress disorder in otherwise well-adjusted personnel must reflect (a) that the disorder and its diagnostic criteria are more descriptive than determinative, (b) that these relationships and their component definitions have tended to be presumptive rather than definitive, (c) that current applications of the diagnostic label seem often to conflict with its derivation, and (d) that these applications and relationships are at best highly questionable when applied to exposures from single events, events with no strong or protracted component of personal jeopardy, and/or events that, while distressing, fall within the range of experiences similarly situated individuals ordinarily and reasonably come to accommodate. In short, both the critical incident as sole precipitator and posttraumatic stress disorder as its direct consequent—while both perhaps widely venerated through ubiquity and repetition—surprisingly but clearly lack solid empirical grounding, while other approaches and constructions of perhaps greater potential utility lie unexplored or tacitly rejected.

APPLICATION TO FIRE AND RESCUE ENTERPRISES

Suggestions that fire and rescue personnel could experience posttraumatic stress disorder through "secondary victimization" began to appear shortly after publication of DSM-III. Much of this discourse appeared first in speculative form, particularly in the nonrefereed trade journals of the industry[33-37] and in trade conferences and convention talks. These quickly grew into proprietary seminars, training programs, and intervention schemes proposing peer intervention teams and similar rubrics hauntingly reminiscent of the crisis intervention movement of 1970s in which several of its principal proponents cut their semiprofessional teeth. One line in particular[33-39] has contended that any and all personnel stand at risk of such decompensation, even from relatively routine exposure; not surprisingly, these arguments have been coupled to equally intemperate contentions that this can only be avoided through application of a specific technique that can be taught by its originator for a fee.[35]

This primarily rhetorical rubric has generated a substantial body of trade literature, including some limited spillover into the applied literature of the counseling trades; virtually all, however, is grounded only in self-citation and oblique references to secondary sources. Some elements, such as Mitchell's "rescue personality" construct, were claimed to have been derived from controlled empirical study,[34,38] but substantive analysis has failed to replicate component claims,[53] and attempts to locate the data from which these profiles were said to have derived has led only to allusions to an unrefereed article[34] with no actual data and to an unpublished study.[38]

More direct study of other major constructs in the critical incident model has raised further doubt regarding both the arguments underlying the enterprise and the labyrinth of roles and relationships forged upon them. Redburn[42,43] reported a systematic study of Sioux City career firefighters exposed to rescue and body recovery operations in a major air crash; despite the presence of virtually all characteristics presented as definitive of the prototypical critical incident,[38,39] no appreciable clinical impact was revealed in a nearly exhaustive sample of participants. Moreover, following an argument grounded in data reported by Alexander and Wells[1] respecting one of the more macabre body recovery operations on record, the researchers suggested that organizational and operational characteristics of the incident, rather than exposure per se, were the primary determinants of perceived stress and that the perception of personal and organizational competency in the conduct of the response was the key determinant of resolution.

Alexander and Wells[1] further noted that, compared to their own baseline data (serendipitously available from testing just prior to the incident) rather than to nonspecific population norms, police officers exposed to an ostensibly traumatizing experience in fact showed actual improvement in measures of postincident affect and functioning. These improvements took place absent any organized psychological intervention, a finding mirrored in Redburn's report that the approximately 40% of the Sioux City firefighters who participated in postincident debriefing showed no clinically significant superiority in their resolutions and, in fact, showed a small but statistically significant deficit compared to those who declined or deferred such participation. While these findings likely represent the consequences of self-selection, the self-veneration of process and technique among adherents and their calls for mandatory debriefing are clearly suspect in light of this pervasive and seemingly reliable noneffect when broader samples are taken and specific symptom indices measured.

ALTERNATIVE CONSTRUCTIONS OF OCCUPATIONAL STRESS AND STRESSORS

Taylor[47] posited more than a decade ago that cognitive adaptation to stressful events might be more effectively characterized from a salutogenic, developmental perspective rather than being presumed pathognomonic of psychopathological risk. A rigorous and productive line of empirical research emerging since that time[49,50] has affirmed and refined these propositions, and they have begun to meld into an alternative, more discernable, and testable construction of the operation of such stressors in both individual and collective adjustment to negative events.[48] More importantly, perhaps, it reflects the longstanding conventional wisdom that adversity is not uniformly threatening but can provide the challenge from which character and resilience are built.

A preliminary study[53] specifically of successful career paramedics lent an interesting partitioning to the question of incident impact. The primary factor influencing perceptions of stressfulness in a series of hypothetical calls hinged not on the manifest content or even on patient outcome but was rather driven by perceptions of personal and organizational success in the address of component evolutions. This was even further highlighted in a thorough study of more than 200 firefighter/paramedics and firefighter/EMTs in the state of Washington,[5] wherein past critical incidents ranked well below such obvious concerns as sleep disruption, wages and benefits, labor/management issues, personal safety, equipment, job skill concerns, and family/financial strains as elements of perceived stress; indeed, past critical incidents failed to significantly enter regression equations predicting job satisfaction or morale for paramedic firefighters and barely achieved significance for firefighter/EMTs.

Motowidlo et al.[40] offered an interactive schematic suggesting how such factors might interact to result in perceptions of occupational stress and to influence attitudinal

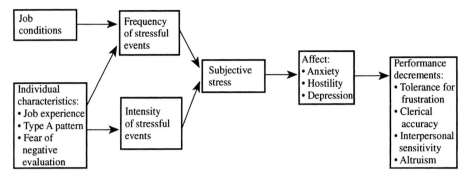

FIGURE 1. Schematic model of occupational stress and workplace impacts. (From Motowidlo SJ, Packard JS, Manning MR: Occupational stress: Its causes and consequences for job performance. J Appl Psychol 71:618–629, 1986; with permission.)

and behavioral concomitants of those perceptions (Fig. 1). Job conditions provide one set of influencing factors, but certain personal characteristics, especially job experience, type A characteristics, and fear of negative evaluation, also strongly influence how events in one's working life will be experienced and interpreted. These factors interact at any given juncture with both the frequency and the intensity of stressful events to influence the subjective reactions of the worker and the affective experience of the events and their contexts; these influences in sequence affect aspects of job performance that then translate readily to behavioral and attitudinal conditions colloquially associated with monikers like burnout. Once again, however, it is not some specific and conceptually isolated set of critical incidents posited to determine the most salient links in this process; it is the backdrop of daily hassles said in fact to make "the big one" loom so large.

Exposure to critical incidents is not only unavoidable in this line of work, it is in fact the essence of the enterprise and provides one of its primary vehicles for effective reward.[16] While the organization may be argued to hold certain responsibilities to ensure that personnel are adequately prepared, equipped, deployed, and configured for effective response and for ensuring that the impact of equivocal events is effectively addressed in organizational and operational review,[17,19] individual decisions, actions, patterns, and responses are also highly determinative of adjustment. Preparation in these aspects of performance must interact with organizational preparedness on a daily basis to ensure that both the exceptional stressors inevitable in the occupation and the routine strains of personal and organizational living that define their contexts are kept in a reasonable balance that presents challenge, generates responsiveness, and encourages personal and professional growth.

ORGANIZATIONAL PREPARATION AND RESPONSE SCHEMATA

The cornerstone of the critical incident model has been the embodiment of its social movement features in the critical incident stress debriefing (CISD) team, generally construed[33,38,39] as a separately chartered volunteer collegium, again strikingly reminiscent in both structure and tone to the crisis intervention centers and hotlines of the early 1970s.[11] These configurations, however, have been fraught with the same peculiar issues that plagued that movement throughout its decade or so of longevity.

Basically, such teams are composed of self-selected personnel who have declared interest in this enterprise and who have attended at least two consecutive days of training, either directly from the originator of this approach or from an approved training program (essentially, from a franchisee); this has now grown to include an advanced two-day program, a peer counseling program, and a family support program, each similarly distinguished by its fiduciary relationship to a primary franchising agent. While attendance at these programs is declared as mandatory if one is to provide "the right kind of help,"[10,35] there are no requirements for examinations or verifications of skill or competence connected to these enterprises; nor is there any evidence of adequacy for these approaches measured against any objective outcome standard; nor do any empirical data provide comparative outcome information that would justify the intemperate claims, both explicit[35] and implied,[10] that other approaches are somehow harmful.

Indeed, much as with the crisis centers studied by Echterling and Wylie[11] a decade and a half back, the politics of the social movement have begun to raise questions and conflicts of their own, both within these teams and their affiliates, between these teams and the public safety organizations whose employees are the intended recipients of their ministrations and between the movement and the established loci of research, practice, and training within the broader academic and psychological communities.[52] Many of these matters stem from the subprofessional or paraprofessional nature of much of this enterprise and from the lack of clear definition regarding roles, relationships, boundaries, and expectations that often accompany enterprises that choose to remain marginal to the established professional arenas, presumably to preserve levels of influence and deference that subprofessional and paraprofessional leaders see as threatened in circles where formal training and refereed research have displaced charisma and ad vericundium authority as the harbingers of status and standing.[11]

Gist and Taylor[17] outlined a series of strategies from an organizational perspective that, they argued, should be considered before, during, and after particularly stressful operations to ensure maximum resiliency for both individuals and their work groups. Rather than proposing any elaborate structure of separately enfranchised peer counselors and therapists as the focal point for organizational preparedness and response, they focused on the incorporation of information and practices into the existing organizational relationships surrounding management, command, supervision, and human resource support, and on the addition of skills and resources to support such roles and interactions where deficits might exist. Moreover, they specifically advocated an approach of empowerment of daily activities and responses over remedial intervention—essentially, opting for flood control engineering rather than sandbagging operations.[20]

Woodall[52] further reported detailed evaluation data from the preliminary presentation of a pilot program designed to identify and present the essential components of an organizationally integrated, theoretically grounded approach to the various component issues identified through a thorough review of the literature of a variety of related fields and disciplines.[18] The first day of the two-day program presented modules relating to the components of daily and organizational functioning that provide the foundation for strong performance and resilience. Those components, their rationales, and the routes of address taken are briefly summarized below.

Basic overview of occupational stress from social, organizational, and community perspectives. This module used case studies, literature review, and operational analyses to outline the premises and arguments providing the foundation for the approach.

Establishment of baseline adaptation strategies. Sound understanding of the nature of occupational stress as it relates to a fire service career is central to its effective address, especially where characterization of stressors as loss, threat, or challenge[30] may be central to both nature and efficacy of the resolution sought. This component (routinely presented in cadet training and as EMT recertification training) reviewed applications of the model suggested by Motowidlo et al.[40] and discussed information and actions to help mediate impacts at each point in the schematic.

Building and working within an effective management climate. It is the daily interactions and issues of the workplace (sometimes called strain) that largely determine the impact of any given stressor.[5,51] This presentation discussed leadership as seen from the working and operating levels of the organization and centered on team building and effective supervision.

Controlling critical incident stress (CIS) through incident command system (ICS). This module, again routinely presented at the cadet level, sought to prepare personnel to work effectively within the Incident Management System built to ensure focused, safe, and effective partitioning of an incident's components and to provide effective address of capacity, capability, and accountability elements necessary to yield effective response.

Family, peer, and professional support systems. This presentation explored the importance of good boundaries and discussed the nature and utilization of such human resource support functions as the Employee Assistance Program and agency chaplains to assist with the various problems in living that occur in any career, as well as those made more prominent by the unique nature of fire and rescue service.

The second day of the program centered on specific applications and techniques. The group process model defining the formal postincident exercise per se stands in many respects as parallel to if not functionally indistinguishable from Mitchell's debriefing format, reflecting essentially a specific application of a general group counseling format diffused throughout the various counseling professions for many years.[8] The placement and conduct of that exercise, its centrality to the overall program, and the parameters surrounding its utilization, however, stand radically juxtaposed between the two approaches.

Mitchell's paradigm portrays the group process as the focal intervention, essential for the prevention of individual decompensation resultant of personal traumatization; the alternative tested by Woodall affords a much less critical role to specific intervention techniques, being organized instead along a continuum from the least formal, structured, and intrusive approaches (given declarative precedence) to the most visible and structured interventions (reserved only for occasions where specific purposes are served in terms of ritual or process through imposition of extraordinary measures). Accordingly, where the Mitchell techniques stand protocol-driven and mechanistic in their description and application, the alternative system promotes instead fluid adaptation of a range of process and provider options to the specific needs of the instant circumstance, favoring always the least formal, least intrusive, and most conservative of the available options.

PROFESSIONAL ISSUES, CLIENT RELATIONSHIPS, AND ETHICAL DILEMMAS

Another major issue nested within choices regarding intervention schemata has to do with client relationships, legal and ethical responsibilities, and definition of roles. Social movements often hold a particular disdain for these sometimes subtle and often prickly questions, purporting that helping those perceived to be in need should simply supersede such abstractions and often tossing glaring ad hominem aspersions toward any who might deign to argue otherwise. Yet these boundaries and definitions can become particularly difficult where employee/employer relationships must be entered into the matrices of roles, responsibilities, and relationships, and can hold very serious consequences if not appropriately managed.

Specific identification of client relationships for CISD teams and their interventions has proven elusive at best. Although identification of the client relationship is arguably the most fundamental question in any professional encounter and is absolutely critical to any meaningful approach to such bedrock matters as informed consent or outcome assessment, this crucial issue finds no satisfactory address in the rubrics of the CISD movement. Indeed, it is commonplace to hear both arguments that debriefing should be mandatory for certain events and to hear reports that indicate, usually quite proudly, that such measures were in fact implemented.[22]

The ethical dilemma here regarding client relationships and informed consent should be immediately self-announcing, and the fact that it lies unaddressed becomes deeply disturbing, especially to the extent that it may indicate failure even to recognize the problem among the practitioners of self-help movements.[9,26,41] The questions of where and in whom responsibility and authority for matters impacting individual and organizational welfare shall be vested, and the standards for the execution and evaluation of the various decisions, actions, and interventions undertaken toward those ends are also elements critical to ethical and effective professional relationships and are matters that seem to receive more evasion than clarification in the standard protocols and prescriptions of the CISD enterprises, again leaving wide areas for potential conflict, miscommunication, and liability.

Mitchell's approach to the matter[38] envisions a volunteer team of peer responders and counselors, hosted and supported by some specific agency but under its own direction and operating under its own protocols and structure. This is again strongly reminiscent of the semiamorphous ad hocracy but paradoxically rigid hierarchies described by Echterling and Wylie[11] for crisis hotlines or by Kaminer[26] for recovery groups, Peele[41] for addiction groups, or Dawes[9] for a variety of self-help groups and self-proclaimed therapies, movements, and seers. Whether their ministrations are in fact organizational interventions geared toward preservation of work team integrity or group interventions geared toward maintaining psychic integrity of some collection of individuals happenstantially related through propinquity to a given event is not entirely clear, although operational suggestions and reported practices, which include simultaneous address of representatives from differing agencies and professions who may have experienced duty in the same basic event,[33,38] strongly suggest the latter. Either way, however, the emphasis on prevention of posttraumatic stress disorder and other presumably consequent psychiatric maladies clearly defines this application as a therapeutic enterprise, despite whatever dodges one might endeavor through attempts to finesse the process outside the realm of regulated activity through semantic manipulation. Accordingly, any such application mandates appropriate licensure, informed consent, clear definition of the client relationship, and full documentation on an individual basis of the services rendered and results derived. None of these factors, however, have been routinely exercised in protocols we have reviewed, and some specifically eschewed.[37,38]

Perhaps even greater pause should be taken from an increasing tendency for CISD teams and their members to extend their services into the civilian population, particularly when those contacts initiate under the auspices of an emergency response provider agency or other governmental entity. The client relationship established through 911 contact clearly does not extend to psychological services of any more broadly therapeutic nature, nor do immunities and protections (or liability coverages) necessarily apply to such ministrations, especially beyond crisis intervention assistance with propinquity in both time and location to the precipitating emergency event. The potential for therapeutic misadventure, including iatrogenic consequences of such encounters, is not inconsequential and merits serious consideration of the ethical implications of "inflicting a painful risk intervention of unproven effectiveness to prevent something that may not happen."[21]

PURSUING A MORE REASONED COURSE

Some resolution may be gained through separating these functions into clear and distinct areas of responsibility and by defining authority and relationships pertinent to each. One model (Community Response Associates, 1995; see Appendix 1) separates these functions into three distinct areas:

Employee and occupational issues, wherein the agency's Employee Assistance Program holds a defined client relationship to the individual employee for assistance with individual issues on or off the job. The EAP is contracted by the agency to pursue such assistance for the explicit purpose of maintaining individual employee functioning and is accountable to the agency for its accomplishment of that end through its confidential assistance to individuals and families for daily hassles as much as for major crises.

Organizational consultation resources are separately contracted to assist the agency and its officers and management in the overall development and maintenance of a positive work climate; it is under this aegis that "critical incident response" is found. The express purpose is distinct from any individual resolutions (the province of the EAP relationship); it is geared instead toward review of the impact of the event on the work team for the distinct purpose of maintaining work team integrity and cohesion. While this is an obvious precursor for workplace and occupational adjustment in the individual, the client relationship belongs here to the organization rather than to the individual employee (whether addressed singularly or in a group context)—indeed, for a single provider or provider agency to attempt to address both sides of this equation yields an immediate conflict of interests!

Services to citizens and customers are provided as social and instrumental support concomitant to operations for which the agency is legally enfranchised and specifically do not extend to—nor should they attempt to mimic or emulate—therapeutic or other direct service relationships beyond the boundaries of triage, referral, and immediate palliative actions. Such initiatives do, however, intersect with any number of nontherapeutic customer service activities[14,15] that seek to help those with whom the agency comes in contact to rebound from their losses and rebuild their community presence in the wake of their emergency incidents. Here, the client relationship is between the agency and its citizen/customer and specifically does not extend or aver to any individual agent, provider, or referee beyond the circumscribed boundaries of the agency's enfranchised services.

Perhaps more significantly, the Standard Service Model for critical incident response (Community Response Associates, 1995; see Appendix 2) is designed to specify a framework for capacity, capability, and accountability at each level of the organization (employee, officer, agency) and thereby to encourage preparation for and execution of an effective systems approach to mitigation, mediation, and remedial address of occupational stressors and occupational strain. This approach specifically eschews the designation of any parallel structure, separate social movement, or externally driven or derived protocols in favor of incorporation of resources, skill, and information to empower the standard actions and interactions inherent in the organization and its operations to effectively address the various components of occupational stress in the workplace. Accordingly, much emphasis is placed on personal resilience, effective supervision, and accountability within management and command systems to ensure that the outcomes sought are supported throughout the daily workings of the agency as well as thoroughly incorporated into the core operations of its response system for major stressful events.

COMPARATIVE TESTS OF MODELS

Woodall's evaluation study[52] sought direct comparison of an organizationally based model[18] against the dominant Mitchell model among a group of law enforcement, fire service, EAP, and education professionals directly acquainted with the latter. Participants reported the program's objectives as highly salient to workplace goals and showed strong preference for its construction, strategies, and grounding. That preference was nearly universal among career fire service participants, for whose organizations the processes were specifically derived; further pilots reiterate these findings.

Despite this, Mitchell adherents continue to offer sweeping but unsubstantiated claims of unique benefit from their approach while suggesting that other approaches are useless or even harmful.[10,35] The only support for these contentions comes from seemingly gratuitous self-citations, which evaporate without foundation in empirical assessment. A single reported study[43a] without objective metric or control has sometimes been cited but on review has proved almost indistinct in form or format from customer satisfaction surveys in hotels or restaurants: it asked only respondents' "liking" of the experience while taking no objective measure of symptoms, resolution, or comparative impact against either alternative treatments or no treatment at all.

Still another recent twist was Dernocoeur's offering[10] of an Australian bank's preliminary and limited report on its internal post-holdup program as definitive evidence to support the sweeping claim that "Research shows that with appropriate critical incident stress intervention, nearly all workers regain their career confidence in a shorter time frame and with less residual devastation than those who do not receive intervention."[10] However, examination of this article showed no statistical tests of differences, no systematic controls or specific symptom or resolution indices, and several clear limitations that would prohibit most any generalization much less any sweeping veneration of the specific CISD approach. Indeed, the article notes first that a major effect of the change was to shift personnel away from general practice physicians who, among other noted limitations, "tended to give automatic extended periods of sick leave"[29] and toward a psychological program formed in part to limit such sick leave use and attendant compensations; it then reports reductions in sick leave and compensation as its sole outcome variables. While this may show that the program met its economic goals, it provides no evidence that "critical incident stress intervention" (a term and approach nowhere cited in the original article) had any measured or measurable effect on "career confidence" or "residual devastation."

HOUSES OF CARDS AND EMPEROR'S ROBES

Dawes[9] noted well the capacity of popularized self-help enterprises to blur the boundaries between the limited and conservatively cast knowledge of science and the aggrandized speculations of those who would wrap their pronouncements in the cloak of presumed scientific expertise, whether to advance their cause or simply to pad their wallets. The fire service, it seems, may be proven unusually vulnerable to this sort of manipulation through the intersection of several significant factors.

1. The fire service has traditionally been an insular enterprise, both socially and occupationally. This implies both that information from outside the enterprise may often fail to enter its exchanges and that, once information and practices become accepted within its boundaries, they may become difficult to challenge.

2. The fire service has traditionally sought technique over theory. The correct way to do something has often been determined more by the activities attendant to its execution than by understanding or application of the principles driving those techniques. This complicates any process of refinement or improvement and can seriously inhibit the progress of more major paradigm shifts.

3. The fire service has accordingly valued training over education and has specifically valued more narrow technical programs over the broad and critical exposures of a liberal education. Again, diversity of viewpoints and critical examination of underlying

principles prove ever more critical as the industry shifts from a basis in technology to an information and service enterprise, and they are crucial to continuous refinement of theory and process.

4. Critical shifts in the nature of the industry have been generationally distinct. The service the current generation is inheriting is quite materially different than that of even the immediately preceding generation[14,19] and has attracted distinctly different types of individuals into the enterprise. One key element has been the shifting of its primary focus from technological actions primarily involving things to informational and direct human service exchanges primarily involving people. This has, in turn, both accelerated and intensified certain challenges in career development that have always accompanied interaction with human loss and suffering.

The currently emerging enterprise, however, has taken these from episodic moments nested at intervals within a career centered on otherwise less personalized activities and interventions and has made them daily fare at the very vortex of the service. The demand for some assistance in these now more central and much more critical adjustments has tended to come from the entering generation and has tended to attract somewhat less engagement from officers whose own experience and value systems were built in what can only be described as a materially different enterprise. No circumstance could be more ripe for a movement driven by "grass roots evangelism," especially when martyrdom is effectively cast as a subliminal but central theme.

The biggest limitation may be that such movements are less likely to attract those who might provide strong models of resiliency and hence those most likely to promote in others solid transition and sound developmental perspective; they too often instead attract those whose own adjustment has been marginal and, hence, frequently yield forums for those who would alleviate their own disequilibrium in part by imputing it to others at vulnerable junctures.[26]

5. The fire service has traditionally treated human resource and management issues as ancillary to its primary mission rather than as critical and central to its execution of that mission and integral to its operations. While technological industries have historically invested first in the equipment that does the job and have then hired personnel to operate that equipment, information and service enterprises invest first in the personnel who deliver the service and then acquire equipment and hardware to make them more effective in that delivery. The difference can be stated simply as a variant of the old canard, "Which came first?" In this case, though, it's the *staff* or the *stuff*.

The fire service has long organized its staff and operations around the stuff, and its rubrics still reflect that paradigm: personnel are assigned to equipment and called by its moniker (e.g., Engine 12, Truck 6); operations are similarly conceptualized by what carried the staff to the scene rather than by the range of skills the staff carried with them. Accordingly, though, the emphasis in officer selection and development has remained technical while the responsibility has become more managerial and developmental. In too many agencies, it seems, such practices tend to yield promotions that effectively deprive the agency of a good nozzle man while yielding a mediocre supervisor.

Traditional stress management and stress intervention programs have failed to integrate practices and principles into the core transactions of management, command, supervision, and related human resource programs and have indeed attempted to claim and defend oddly separate turf.* This may certainly serve to protect the proprietary interests of the movement's self-anointed leadership[9] and to maintain the in-group status of its local

---

*Mitchell and Bray[38] discuss "lead agency" obligations to the "multiagency, multijurisdictional" team. Note, however, the contention that the team must effectively operate independently of management by the agency or agencies expected to support it—backed, once again, by citation only of a nondata, unrefereed piece in a trade journal of limited circulation.

membership while serving any personal needs for vindication or veneration,[26] but it also compounds informational insulation and again promotes groupthink characteristics.[25]

This promotes, in effect, the interests of the movement itself, but paradoxically does so at the expense of the arguably more effective approach to mediation and mitigation—effective integration into routine exchanges that work to make stress impact ever more a nonissue at remedial levels. This may partly reflect a paradigmatic disparity concerning whether stress exposure must always be considered harmful and hence always demand remedial response*[35,37,38] or whether such events should be seen as challenges to systems and responders, which, if properly managed and addressed, have as much potential to enhance and empower as to inhibit or harm.[1,17,20,42,43,52] Again, however, the former position is based primarily on assertion while the latter position represents convergence of significant empirical study.

6. The vehicles for information dissemination in the fire service have not developed systematic means to ensure quality of the data, arguments, or positions espoused. Unlike academic and scientific literature and conferences, the trade journals and trade conferences that provide the nearly exclusive information channels to the working fire service are primarily journalistic and/or proprietary and may tend to be effectively governed by issues of subscription even before issues of accuracy or soundness of data and argument. The system of blind referees employed in the former venues, despite its faults and shortcomings, provides an important vehicle by which to ensure that pseudoscience, pontification, mercenary manipulation, and self-aggrandizement are kept to the greatest extent possible from using the cloak of science to advance their positions; an industry shifting its foundation to information and service faces a significant challenge to ensure that it develops some similar method to preserve its informational integrity—both for its own protection and, even more critically, for the protection of the public it serves.

MAKING OUR EFFORTS EFFECTIVE

Stern offers the social function of science as *"(to) separate common sense from common nonsense and (to) make uncommon sense more common."*[9,46] We have said the same thing another way: our most important lessons in dealing with occupational stress may come more from grandma than from graduate school, but our efforts as scientists, practitioners, and educators must be to determine with some reasonable certainty just what it was that made grandma so astute. Gist[13] made these observations in his report to Sioux City firefighters on Redburn's[42] research there:

> We do not mean to say that the debriefing process should be discarded, any more than we would offer any other contention that babies should be thrown out with bath water. What we are finding, however, is that the things that work best are those which are closest to our roots, and that we may fare better by making our existing approaches more effective than by attempting to radically shift the focus . . . as I am often known to put it, the fire service got along for two centuries without me; I've got to assume that they were doing a few things right. When we look at what those things are, they suggest not that firefighters are vulnerable to stress disorders, but rather that they are extraordinarily stress resistant! By exploring the factors which enhance that resistance, we have so far concluded that strong attention to management, command, and supervision coupled with emphasis on training and physical conditioning gives us the best preparation; we have also concluded, interestingly enough, that the less intrusive or formal the postincident response, the healthier the outcome. In effect, we save the major intervention for the major event, but build it as a simple extension of our ordinary activities—much like ICS, we allow it to grow and contract with the demands of the incident, but absorb it into the daily system to make it effective when needed most.

---

*These works and other reports of the Mitchell CISD model consistently rely on lists of "critical incidents" that would always demand intervention; no empirical grounding for these lists, however, has been offered.

Gist and Woodall[20] placed this into perspective by recalling the remarks of a retired battalion chief, introduced to his agency's "stress guy" and told how it was a shame that such a fellow hadn't been available when the retiree had been seriously injured on the job some years before. The retiree retorted that there certainly had been such folks: "We used to call them captains and chiefs."

The point best made and best remembered remains, perhaps, simply that psychologists can certainly help captains and chiefs to do their work more effectively, but it remains the work of captains and chiefs to do. Two seemingly paradoxical invectives must be balanced to make that relationship effective. First, of course, we must recall that good fences make good neighbors—psychologists who want to play firefighter and firefighters who want to play psychologist are both dangers to self and others; social movement gurus who grandiosely overstate their quite limited claims to either status as if a resolution to the boundaries between them are a danger to everyone. Secondly, though, we must learn to talk constructively over the fence if we are to have science supersede shamanism and showmanship in building the health, safety, and service of the enterprise whose missions we proudly share.

REFERENCES

1. Alexander DA, Wells A: Reactions of police officers to body handling after a major disaster: A before and after comparison. Br J Psychiatry 159:547–555, 1991.
2. American Psychiatric Association: Diagnostic and Statistical Manual. 2nd ed. Washington, DC, American Psychiatric Association, 1968.
3. American Psychiatric Association: Diagnostic and Statistical Manual. 3rd ed. Washington, DC, American Psychiatric Association, 1980.
4. American Psychiatric Association: Diagnostic and Statistical Manual. 3rd ed. rev. Washington, DC, American Psychiatric Association, 1987.
5. Beaton RD, Murphy SA: Sources of occupational stress among firefighter/EMTs and firefighter/paramedics and correlations with job-related outcomes. Prehospital Dis Med 8:140–150, 1993.
6. Breslau N, Davis GC: Posttraumatic stress disorder: The stressor criterion. J Nerv Ment Dis 175:255–264, 1987.
7. Cook JD, Bickman L: Social support and psychological symptomatology following a natural disaster. J Traum Stress 3:541–556, 1990.
8. Corey G: Theory and Practice of Group Counseling. 4th ed. Pacific Grove, CA, Brooks/Cole, 1995.
9. Dawes RM: House of Cards: Psychology and Psychotherapy Built on Myth. New York, Free Press, 1994.
10. Dernocoeur K: Are we getting the help we need? J Emerg Med Serv 20(8):30–36, 1995.
11. Echterling L, Wylie ML: Crisis centers: A social movement perspective. J Community Psychol 9:342–346, 1981.
12. Folkman S: Personal control and stress and coping processes: A theoretical analysis. J Pers Soc Psychol 46:839–852, 1984.
13. Gist R: Memorandum to Sioux City firefighters summarizing research findings. July 31, 1992.
14. Gist R, Brisbin SR: It's never quite what you thought it was: Data trends and paradigm shifts in the contemporary fire service [manuscript].
15. Gist R, Gilchrist LP: Tasty, friendly, pretty, and packaged: Beyond "hot & quick" in customer service [manuscript].
16. Gist R, Obadal R: How to Stay Alive and Prosper in EMS. Overload Park, KS, EMT Recertification Program, Center for Continuing Professional Education, Johnson County Community College, 1994.
17. Gist R, Taylor VH: Coworkers and their families. West Virginia Trooper 7(2):93–99, 1992.
18. Gist R, Taylor R, Woodall SJ, Magenheimer LK: Personal, Organizational, and Agency Development: The Psychological Dimension. Phoenix, St. Luke's Behavioral Health System/Phoenix Fire Department, 1994.
19. Gist R, Tvedten J: A human systems look at fire service organization [manuscript].
20. Gist R, Woodall SJ: ". . . and then you do the hokey-pokey, and you turn yourself about" [manuscript].
21. Grollmes EE: Post disaster: Preserving and/or recovering the self. Dis Manage 4(2):150–156, 1992.
22. Hansen J: The Alfred P. Murrah Federal Building bombing. Presented at the 13th Biennial Symposium on Occupational Health and Hazards of the Fire Service, John P. Redmond Foundation of the International Association of Fire Fighters, San Francisco, August 28, 1995.
23. Helmreich WB: Against All Odds: Holocaust Survivors and the Successful Lives They Led. New York, Simon & Schuster, 1992.

24. Hoffer E: The True Believer: Thoughts on the Nature of Mass Movements. New York, Harper & Row, 1951.
25. Janis IL: Groupthink: Psychological Studies of Policy Decisions and Fiascoes. 2nd ed. Boston, Houghton-Mifflin, 1982.
26. Kaminer W: I'm Dysfunctional, You're Dysfunctional: The Recovery Movement and Other Self-Help Fashions. Reading, MA, Addison-Wesley, 1992.
27. King DW, King LA: Validity issues in research on Viet Nam veteran adjustment. Psychol Bull 109: 104–124, 1991.
28. Krantz L: Jobs Rated Almanac, Two. Mahwah, NJ: World Almanac, 1992.
29. Leeman-Connely M: After a violent robbery. Criminology Australia, April/May: 4–6, 1990.
30. McCrae RR: Situational determinants of coping responses: Loss, threat, and challenge. J Pers Soc Psychol 46:919–928, 1984.
31. McFarlane AC: Relationship between psychiatric impairment and natural disaster: The role of distress. Psychological Med 18:129–139, 1988.
32. McFarlane AC: The aetiology of post-traumatic morbidity: Predisposing, precipitating, and perpetuating factors. Br J Psychiatry 154:221–228, 1989.
33. Mitchell JT: Development and functions of a critical incident stress debriefing team. J Emerg Med Serv 13(12):42–46, 1988.
34. Mitchell JT: Living dangerously: Why some firefighters take risks on the job. Firehouse 11(8):50–51, 63, 1986.
35. Mitchell JT: Protecting your people from critical incident stress. Fire Chief 36(5):61–64, 1992.
36. Mitchell JT: The history, status, and future of critical incident stress debriefing. J Emerg Med Serv 13(11):49–52, 1988.
37. Mitchell JT: When disaster strikes . . . The critical incident stress debriefing process. J Emerg Med Serv 8(1):36–39, 1983.
38. Mitchell JT, Bray G: Emergency Services Stress. Englewood Cliffs, NJ, Brady, 1990.
39. Mitchell JT, Everly GS: Critical Incident Stress Debriefing: An Operations Manual for the Prevention of Traumatic Stress Among Emergency Services and Disaster Workers. Ellicott City, MD, Chevron Publishing, 1993.
40. Motowidlo SJ, Packard JS, Manning MR: Occupational stress: Its causes and consequences for job performance. J Appl Psychol 71:618–629, 1986.
41. Peele S: The Diseasing of America: Addiction Treatment out of Control. Lexington, MA, Lexington Books, 1989.
42. Redburn, BG: Disaster and rescue: Worker effects and coping strategies [doctoral dissertation]. Kansas City, MO, University of Missouri, 1992.
43. Redburn BG, Gensheimer LK, Gist R: Disaster aftermath: Social support among resilient rescue workers. Presented at the Fourth Biennial Conference on Community Research and Action, Society for Community Research and Action, Williamsburg, VA, June 1993.
43a. Robinson RC, Mitchell JT: Evaluation of psychological debriefings. J Traum Stress 6:367–378, 1993.
44. Schroeder DH, Costa PT: Influence of life events on physical illness: Substantive effects or methodological flaws? J Person Soc Psychol 46:853–863, 1984.
45. Solomon SD, Canino GJ: Appropriateness of the DSM-III-R criteria for posttraumatic stress disorder. Compr Psychiatry 31:227–237, 1990.
46. Stern PC: A second environmental science: Human-environment interactions. Science 260:1997–1999, 1993.
47. Taylor SE: Adjusting to threatening events: A theory of cognitive adaptation. Am Psychologist 38:1161–1173, 1983.
48. Taylor SE: Asymmetrical effects of positive and negative events: The mobilization-minimization hypothesis. Psychol Bull 110:67–85, 1991.
49. Taylor SE, Brown J D: Illusion and well-being: A social psychological perspective on mental health. Psychol Bull 103:193–211, 1988.
50. Taylor SE, Lobel M: Social comparison activity under threat: Downward evaluation and upward contacts. Psychol Rev 96:569–575, 1989.
51. Wagner BM, Compas BE, Howell DC: Daily and major life events: A test of an integrative model of psychosocial stress. Am J Community Psychol 16:189–205, 1988.
52. Woodall SJ: Personal, organizational, and agency development: The psychological dimension. A closer examination of critical incident stress management. Emmitsburg, MD, National Fire Academy Executive Fire Officer program, 1994.
53. Wright RM: Any fool can face a crisis: A look at the daily issues that make an incident critical. In Gist R, (moderator): New Information, New Approaches, New Ideas. Overland Park, KS, Center for Continuing Professional Education, Johnson County Community College, 1993.

Appendix 1

Standard Service Model

Policy Statement and Parameters

PSYCHOLOGICAL AND ORGANIZATIONAL ASSISTANCE

The contemporary fire service agency must recognize that its mission and its operations include many elements which mandate the prudent and effective application of psychological, sociological, and behavioral principles. These principles apply at a variety of levels, and include services to assist personnel in the resolution of personal, family, and occupational issues which may interfere with their capacity for optimal functioning; services to help identify and promote the effective address of social and psychological impacts among those who receive our services; and services to assist our organization in the selection, development, and evaluation of its personnel, policies, and procedures. This policy defines each level of service; establishes the capacities, capabilities, and accountabilities required to establish and deliver an effective standard of service in each domain; and provides standard procedures for the implementation of effective service at each level of operation.

Personal, Family, and Occupational Issues

The agency provides for its members and their families access to a comprehensive Employee Assistance Program as a benefit of employment. This program is responsible for providing confidential assistance to any member respecting personal or family issues, and to provide either direct counseling services and/or prompt referral to appropriate service providers to assist the member or eligible dependent in the resolution of such issues. The description of services and access procedures for the agency's current contractor is attached to the appropriate procedural memorandum (see Employee Assistance Program) and incorporated by reference.

The agency may also refer an employee to the Employee Assistance Program for assistance in the remedy of specific occupational problems, and may specifically prescribe the service as a resource for any employee whose personal problems or issues may be adversely affecting his or her work performance. Mandatory referral will come only at the third level of the agency's progressive discipline policy (following verbal and written warnings), and will be accompanied by a specific set of behavioral expectations defined by the supervisor in conjunction with the agency's organizational consultants (see Organizational Consultation below); that referral will include specific description of the problem behaviors or issues, clear specification of the impact of those behaviors or issues, and clear specification of the behaviors expected to reestablish satisfactory performance. Progress reports, limited to referral compliance and assessment of progress on relevant behaviors and issues, will be made to the agency's organizational consultants and will be treated as confidential records; the agency may also prescribe independent evaluation of progress by or through its organizational consultants.

The primary client relationship for the Employee Assistance Program resides between the employee and his or her counselor and, within the limits noted in the preceding paragraph (and as otherwise prescribed by law), all elements of confidentially attendant to such relationships remain afforded the employee.

Services to Citizens and Customers

The social and psychological impact of the events experienced by the citizens we serve forms for them a major element of that experience. Our mission of service and aid to those in need demands that we stand prepared to help them withstand those impacts, and that we endeavor to assist them in locating and utilizing all appropriate resources which might contribute to their optimal recovery. The agency has accordingly developed procedures by which to prepare personnel to provide appropriate assistance, and through which to ensure that resources and responsibility are assigned to its delivery in all pertinent operations (see Customer Services Sector).

These services represent an extension of the citizen's request for service through 911, and are designed specifically to provide support and comfort for those requiring our services, to help minimize negative consequences of the experience for the citizens involved, to begin the identification and mobilization of additional social and community services which might contribute to the recovery and well-being of those we serve, and to provide for any necessary or desirable follow-up contact to ensure that all pertinent services have been identified and secured. While basic crisis intervention and social assessment skills are applied, the agency does not assert or assume expertise in the direct provision of psychological care, and does not itself function as an agent for such delivery, whether on its own or on the part of other organizations or providers. The efforts we provide are specifically defined as the provision of social support and instrumental aid through triage, assessment, referral, and system advocacy in direct extension of our fire, rescue, and EMS purviews and authority.

The agency provides training and consultation for this function through its agreement with Community Response Associates. Sector personnel in the field may request consultation or response through the procedures specified in the appropriate directive (see Customer Service Sector). The client relationship at this level of operation holds between the agency and the citizen or customer, and is restricted specifically to that domain; the services of other agencies may be solicited on behalf of the citizen and offered for the citizen's consideration, but must remain clearly independent of any categorical endorsement or assumption of liability on behalf of the District, its officers, or its agents.

Organization Consultation

Satisfaction of our agency's service mission requires that we develop and hold the capacity, capability, and accountability to deliver an effective standard of service in each aspect of our service delivery system. The agency also maintains resources to assist its management staff in such areas as personnel recruitment, selection, training, development, and evaluation; to provide consultation in the establishment, development, and dissemination of agency policies, procedures, and standard service models; and to provide direct consultation to officers, managers, and supervisors in any matters related to their duties and responsibilities. This will also include assistance to the agency in responding to extraordinary events or conditions in the workplace (see Critical Incident Response). These services are separate and distinct from those of the Employee Assistance Program, and are provided through a separate agreement with Community Response Associates.

Supervisory and management personnel may request consultation regarding any aspect of personnel or organizational performance, and shall specifically request such consultation in the formulation of any mandatory EAP referral (see Personal, Family, and Occupational Issues above; see also Employee Assistance Program) or for assessment of critical workplace events and recommendation of appropriate responses (see Critical Incident Response). The client relationship resides between CRA and the organization, and extends to its managers and supervisors as delegated agents of the organization.

Appendix 2

Standard Service Model

Policy Statement and Parameters

CRITICAL INCIDENT RESPONSE

Fire and rescue personnel confront many difficult circumstances in the course of their careers. They are consistently exposed to high tension, high demand situations where difficult decisions of extraordinary consequence must be made quickly, and where those decisions must sometimes be made without the benefit of complete information or time to consider every possibility. They also work regularly under conditions which contain an inherent element of danger, and which can very quickly turn from the seemingly routine to the potentially deadly.

Exposure to such unsettling experiences demands that the employee be ready to adapt to challenges to his or her view of the world, the career, and his or her relationship to agency, occupation, and community. These adjustments, if taken properly, can contribute to strength, character, and commitment; they also, however, hold the potential to affect the employee and the agency in negative or nonproductive ways. Certain events can hold the capacity to seriously disrupt the ordinary vehicles of social support through which veteran members have traditionally helped those shaken by novel experiences to weather the impact and establish perspective; in these circumstances especially, the agency will take precautionary steps to ensure that the normal coping systems of the organization, both formal and informal, are focused on healthy and productive adaptation.

Positive adjustment to the impact of particular events experienced in the course of one's career is a crucial element in effective employee development, and the agency has established this standard for maintaining the capacity, capability, and accountability for assisting employees in optimal adjustment to the demands of their careers. Specific standards are defined for each level of the organization, and include responsibilities of employees, officers, and the agency in routine as well as extraordinary circumstances.

Purpose

The clearest conclusion from both research and practice regarding critical incident reactions is simple and straightforward: The best preventative is a well managed incident conducted by a well managed agency; this requires well trained, well conditioned, well supervised, and well grounded personnel working together to achieve a set of well defined outcomes through well prepared and well rehearsed operations and procedures. The agency works toward establishing this platform through three basic functional areas:

    a. Management of the agency will be based on providing a stable but dynamic operating climate in which the organization can maintain a clear focus on a defined mission, determine and define standard service outcomes for each element of that mission, develop and execute standard procedures through which to achieve those core outcomes, and establish the capacities, capabilities, and accountabilities needed to ensure the satisfaction of its mission through the consistent delivery of competent, quality service.

    b. Command procedures within the agency will be established and employed in a consistent and uniform fashion to ensure that outcomes and procedures are activated, coordinated, and focused in the direct delivery phases of each service episode.

    c. Supervision practices will be focused on progressive employee development and continuous quality improvement, with a strong and positive focus at the level of direct service delivery; the primary objective shall be to ensure that practices and actions deliver standard outcomes through safe, consistent, and effective operations.

All aspects of this system hold in common an express focus on creating an environment in which competent and committed professionals can function effectively in demanding, highly charged, and highly changeable situations. The agency is therefore strongly committed to ensuring that all necessary resources are provided to support such performance, and that all means necessary to ensure optimum employee growth, adaptation, and adjustment are available when needed.

Capacity, capability, and accountability for critical incident preparation and response have been established at each level of responsibility to ensure that effective employee support is provided under any circumstance in which reaction to or adjustment from work place demands may inhibit individual or work team resilience. These procedures are not reserved for rare or atypical events, but should instead be incorporated into all routine aspects of performance.

Employee Responsibilities

The most important determinant of any individual's response to stress, whether in the wake of a single extraordinary event or from the cumulative impact of relatively routine episodes, depends in greatest measure on that individual's personal fitness and resiliency, and on the solidity and effectiveness of his or her immediate support systems. Each member of the agency individually holds the professional responsibility to monitor and maintain his or her personal fitness, both physically and psychologically, to ensure capability to perform adequately under high demand, withstand the impact of major challenge, and rebound from intense demand without undue personal or professional compromise.

Since any individual may experience difficulties from time to time, whether related to personal, family, or occupational issues, and since any such unresolved issues may compromise that individual's capacity for effective performance and optimal resilience, members also hold the individual and collective professional responsibility to remain attentive to coworkers and to stand ready to support and assist a fellow member through the provision of collegial support and/or through assistance in securing appropriate aid. The organizational climate is also critical to effective functioning and to resilience under duress; each member holds the responsibility to work diligently toward a positive professional atmosphere in which to serve and function, and to maintain positive and professional bearing and demeanor in all interactions with colleagues, customers, and constituents.

The following capacities, capabilities, and accountabilities apply to this level of service:

Capacity

1. The agency will provide for mental health, substance abuse, family counseling, and related treatment services as a component of its health insurance benefits. These shall include coverage for crisis stabilization, inpatient and outpatient services, residential treatment programs for substance abuse, and marriage and family intervention (see Employee Benefits).

2. The agency will provide a comprehensive Employee Assistance Program for members and families. This program will provide at a minimum confidential response to requests for crisis intervention, triage of personal and family systems issues, assessment and referral for substance abuse or related problems, and such additional services as may be needed to assist employees with personal, family, or occupational issues (see Psychological and Organizational Assistance and Employee Assistance Program).

3. The agency will provide identified peer support personnel to assist members outside the formal structures. These individuals can be called upon to provide confidential assistance to members seeking information or support beyond that mandated as a component of individual responsibility, and to strengthen the linkage between informal and formal support structures of the organization.

4. The agency will also establish and support a program of personal fitness and wellness, and will establish specific standards consistent with NFPA health and fitness standards.

Capability

1. Each employee will receive basic instruction in access and use of the Employee Assistance Program. This will include both written and verbal presentation of benefits and access, and will emphasize the confidentiality of contacts; these presentations will also include information

regarding applicable health insurance benefits and their utilization, whether in conjunction with, on referral by, or independently of the EAP. Any change in benefits, procedures, or provider will be preceded by complete information delivered to all members in the same fashion as described above; new employees shall receive this instruction as a component of initial training and orientation.

2. At least two senior members from each shift shall be recruited to serve as peer support personnel and will be provided specific training for this role. These individuals may be contacted informally by any member for confidential assistance, and will be prepared to assist with basic crisis support, exploration of options, and access to resources. They will also be prepared to execute peer support functions related to incident response and review (see Officer Responsibilities, Command Responsibilities, and Management Responsibilities below).

3. The Health and Safety Committee is charged with the development of specific standards for personal fitness, and with recommendation of appropriate programs for their implementation and assessment. These standards shall be consistent with applicable standards of NFPA, and implementation procedures must provide reasonable assurance of adequate compliance both by the agency and by employees; standards and procedures will be reviewed at least annually.

4. Each member will receive instruction in the agency's personal fitness standards, and assistance in the development of his or her personal fitness goals and training plan. The agency will provide such assistance as may be reasonably required to facilitate this aspect of personal planning, including necessary evaluations and consultations (e.g., body fat composition; cholesterol screening; aerobic reserve capacity) as may be required for effective implementation.

5. Each employee will also receive specific instruction regarding occupational stress in fire and rescue, and regarding personal and organizational responsibilities for its address and resolution. This will include both general instruction as a part of initial training and orientation, and specific instruction regarding agency policies, procedures, and resources as a part of inservice development.

ACCOUNTABILITY

1. Executive management is charged with securing by contract a comprehensive Employee Assistance Program, and with the annual review of the program's goals and objectives, the contract through which those goals and objectives are met, and compliance and performance of the program and provider. The agency's organizational consultants may be charged with the design and execution Requests for Proposals, bid review processes, and evaluation studies, subject to review and acceptance by executive management and Board.

2. The Assistant Chief for EMS and Training is charged with initial training of members regarding EAP benefits, health benefits, and fitness standards, and with ensuring subsequent compliance with training requirements established in this policy. This will include incorporation of such instruction, including mental wellness and occupational stress components, into the training and orientation of new personnel.

3. Battalion and division chiefs are charged with the selection of peer support personnel within their areas of responsibility; the Assistant Chief for EMS and Training will be responsible for coordinating initial training, and for subsequent training as indicated. The agency's organizational consultants may be charged with the development of position descriptions, selection standards, and procedures for the recruitment, selection, and review of candidates for these roles; they may also be charged with the design and execution of the precise training curriculum, subject to the approval of the Assistant Chief for EMS and Training and executive management.

4. Each battalion officer, division officer, and station officer is charged with maintaining awareness of and contact with peer support personnel in his or her area of responsibility, and for ensuring that their identities and accessibility are made regularly known to personnel under their charge.

5. Peer support personnel will be accountable for response to requests for assistance within their area of assignment, and from other areas at their discretion. Each is also charged with maintenance of confidentiality of all contacts (within the limits prescribed by law), and with the active solicitation of consultation from EAP and/or agency consultants if and when indicated.

6. Individual employees are charged with compliance with all aspects of policy and procedure, and specifically with compliance with fitness standards, maintenance of personal health and

fitness for duty (both physical and psychological), and for employing any and all appropriate avenues of assistance to affirmatively rectify any problem or situation which might compromise his or her fitness, safety, optimal performance, or resilience. Such compliance is a legitimate element of occupational functioning and professional responsibility, and may be considered in Annual Performance Review, Fitness for Duty, and/or Corrective Discipline (including mandatory EAP referral; see Annual Performance Review, Fitness for Duty, Employee Assistance Program, and Psychological and Organizational Assistance [Personal, Family, and Occupational Issues]).

7. When any incident, event, or experience in the context of the workplace creates for an employee unusual discomfort, distress, or difficulty in adjustment, that employee is charged with seeking the assistance of a company officer, battalion officer, peer support provider, and/or EAP contact. Such contact alone will not form a basis for disciplinary action or judgement regarding fitness for duty, and shall in fact be considered affirmative execution of responsibilities noted under (6) above when initiated in an appropriate and timely manner.

8. Any employee who notes what he or she believes to be such difficulties in the adjustment of a colleague shall make every reasonable and prudent attempt to assist that person in achieving resilient adaptation. That responsibility includes, where safety of the employee of others may be compromised by the colleague's condition or actions, the affirmative responsibility to inform an officer and/or peer support provider so that effective intervention can be attempted.

Officer Responsibilities

Officers are specifically responsible for the effective utilization, development, assessment, and support of personnel under their charge. Essential elements of that responsibility include assistance to employees in addressing and accommodating challenges in their career development, and the recognition and rectification of transitory impediments to effective job functioning, including circumstances which might impede response to or resilience from major challenges, threats, or disruptions. Each officer is expected to maintain an awareness of fitness and resilience factors within his or her employees at any given time, and to provide timely assistance and intervention where indicated to ensure optimal functioning is maintained or restored.

Each officer must be prepared at all times to monitor and evaluate the impact of specific incidents and experiences on effective employee and work team functioning, and to take appropriate measures to ensure resilient, constructive resolution of those experiences after exposure. These latter responsibilities are expected features of effective supervision, are ordinarily integrated into the systematic review of performance and operations which provides for growth and development in both technical and personal aspects of the profession. Additional resources and procedures may be called upon to address more difficult circumstances, and it is the responsibility of each officer to be aware of those resources, assess the need for their application, and initiate any procedures which may be called for in any given circumstance.

Integrity of the work team is a primary officer responsibility, and officers must be particularly attentive to the impact of circumstances and events in the workplace on the capacity and capability for effective team functioning. Officers are expected to take proactive steps to ensure that the impact specific events and incidents may hold for work team integrity anticipated and addressed at the earliest feasible point, and to make full use of appropriate agency resources to effect positive resolution. Specific measures will include direct attention to impact of exposure in company-level review of operations, and may include involvement of peer support personnel, senior staff, and/or agency consultants to provide more structured support. Other appropriate actions and activities (beyond standard elements of operational review) may include:

a. Request for station visit from a peer support responder not directly involved in the incident at point. This can be done as an informal visit, or as a part of a more organized discussion of the event and its implications.

b. Request for assessment and/or assistance through the agency's organizational consultants. This may involve telephone review and recommendations, station visits (ordinarily in conjunction with peer support personnel), or recommendation for structured intervention (i.e., formal debriefing).

c. Request for formal debriefing session. This process involves the structured review of the incident and its impact utilizing both the agency's organizational consultants and selected peer support personnel; representatives of the Employee Assistance Program may also be requested to participate for the express purpose of facilitating any subsequent employee contacts which might emerge. This process focuses on reconstruction of a clear picture of the event, examination of its impact on individuals and work team integrity, anticipatory guidance respecting reactions and resolutions, and identification of practical skills and resources through which to achieve positive adjustment; it is accordingly defined as an organizational (rather than clinical) intervention, and will be treated as an element of operational review precursory to and distinct from review of technical aspects of the incident.

Capacity

1. The agency provides access to its organizational consultants for any officer requesting assistance in the evaluation of employee issues or the formulation of intervention and/or referral options. This resource may be employed in any circumstance in which the officer desires additional input at a professional level, and shall be employed in any circumstance where mandatory referral to the Employee Assistance Program is contemplated as an element of effective resolution (see Psychological and Organizational Assistance).

2. The agency provides a comprehensive Employee Assistance Program which may be offered as a referral for individual employee problems. EAP referral may also be prescriptively employed where specific and defined problems create measurable impediments to effective functioning, and where specific job-related behavior change is sought (see Psychological and Organizational Assistance and Employee Assistance Program).

3. Peer support personnel are provided for informal assistance in the exploration and resolution of employee issues. Officers may recommend that any employee explore this resource regarding any issue. This resource is also intended to enhance the strength of the informal organization in mutual support of our members, and specifically to provide the "veteran to rookie" contacts which have always characterized the positive aspects of career growth and response to challenge in the fire service.

4. The agency's organizational consultants are prepared both to assist officers in evaluation of situations and options, and to provide direct assistance in the execution of formal interventions when requested. These resources may be accessed through any chief officer, and are generally available on a 24-hour basis.

5. The agency will maintain mutual aid protocols for critical incident response. These will specify what assistance may be requested from other agencies, under what circumstances it may be requested, and how such assistance, when requested, shall be employed; these protocols shall also list those agencies from whom such aid may be obtained. Those protocols shall also define under what circumstances such aid may be requested from this agency, and shall define how such requests will be processed and how such assistance shall be deployed.

Capability

1. Officers shall receive systematic training regarding the philosophy and mechanics of effective developmental supervision. That training shall include specific attention to employee development planning, procedures for assessment of both job performance and career development, and techniques for recognition of and intervention with situationally induced impediments to effective work and work team performance.

2. Officer training shall include specific components regarding effective use of peer support personnel, Employee Assistance Program, critical incident response, and organizational consultants. Alterations in any standards and protocols relating to these resources shall be incorporated into inservice instruction for officers and supervisors.

3. Training in command procedures will include specific attention to psychological and psychophysiological aspects of performance, and to applications within the incident command system which promote optimum performance, minimize impact of exposure, and enhance positive resolution and adaptation (see Command Procedures).

Accountability

1. Company officers hold primary accountability for systematic assessment of functional status and service fitness of employees in their charge. This will be a primary component of Performance Review for all company and battalion officers, and will include systematic review of

officer efforts in individual development planning, performance monitoring, and use of appropriate resources for assistance and intervention.

2. Company officers are responsible for systematic review with employees of all company operations. This will specifically include assessment and address of any pertinent issues regarding impact or exposure, and application of any resources and/or procedures necessary to effect optimal resolution and employee growth.

3. Company officers will maintain awareness of impact issues in staging, assignment, and operations within single company operations, and will make all feasible and prudent efforts to regulate and address those issues in a preventive and proactive manner.

4. Battalion and shift officers are responsible for providing proactive support to company officers in the execution of these responsibilities, and will monitor and review the effectiveness of those efforts on a consistent and continuing basis. These will be primary accountabilities in Performance Review at this level.

5. Battalion and shift officers will assume responsibility for coordinating such efforts where multiple company operations are involved. Responsibilities, accountabilities, and procedures established under this section will apply similarly at this level of operation and response.

Agency Responsibilities

Complex operations, including multicompany, multijurisidictional, and/or multiple casualty operations, present particular challenges and demand particular attention in their resolution. Systematic operational review may be impeded unless and until reactions are processed and equilibrium established; this may in turn inhibit effective resolution of the experience at both the individual and work team levels. Proactive measures shall be undertaken to ensure that all resources and procedures necessary to ensure appropriate response are identified and executed.

The Incident Command structure of the agency will specifically address both prevention and remediation in all relevant functional areas to ensure that its operations remain safe, effective, and focused. Specific areas of significance include:

   a. Safety Officer responsibilities will include assurance that stress impacts are not allowed to adversely affect the capacity and capability for safe and effective operations during any incident.

   b. Staging Officer responsibilities shall include adequate briefing and communication of assignment before transmittal to Operations, and provision of all other information and resources necessary for effective commencement of activity.

   c. Sector officers and team leaders in Operations Sector shall ensure that personnel are not extended beyond their optimum exertion and performance level (i.e., allowed to approach exhaustion or sympathetic collapse) before relieved of immediate assignment and transmitted to Rehab.

   d. Rehab Sector responsibilities shall include assessment, remediation, and stabilization of both physical and psychological manifestations of stress and exertion before transmittal to Staging for reassignment or demobilization from the scene.

   e. Liaison officers (including Customer Services Sector) shall ensure that individuals working in these areas appropriately address needs of those outside core operations.

   f. Incident Commander shall ensure that all sectors and responsibilities necessary to ensure execution of these responsibilities are established, as required by the characteristics and circumstances of the specific incident.

Measures taken after termination of operations may include any of the steps outlined under Officer Responsibilities above, and may specifically include mutual aid requests for debriefing or other intervention assistance if:

   a. agency peer response personnel were themselves involved in the incident; or

   b. agency peer response personnel not directly involved in the incident might reasonably be expected to carry impacts of the event (e.g., line of duty death of a coworker; citizen death in operations); or

   c. agency peer response personnel lack direct experience relevant to the circumstances of the event; or

   d. executive management, agency consultants, or responsible officers determine such a request to be advisable in the particular circumstances of the event.

Mutual aid requests will be made only for resources of agencies predetermined by specific agreements, and their operation will be determined by the parameters of those agreements and the protocols of the respondent agency. Resources of clinical provider groups, mental health practitioners, or other ad hoc sources will not be authorized to supplant any established element of these procedures.

Capacity

1. The agency will maintain the capacity to identify and establish all ICS sectors necessary to ensure safe, effective, and competent operation of any complex incident, including all components necessary to minimize negative exposure and promote optimum resiliency and adjustment.

2. The agency will maintain one or more specific mutual aid agreements regarding critical incident response. These agreements will be undertaken only with agencies whose protocols and procedures maintain relative uniformity with the goals, methods, and outcomes established herein for this function within the agency.

3. The agency will also promulgate a standard mutual aid agreement regarding access to its critical incident response capacity by other jurisdictions. This agreement will specify the standard service outcome for the resource, its standard operating procedures, and the accountabilities attendant to its access and delivery; that information will be made available to any agency which might request such assistance.

Capability

1. Incident Command System training for officers will address competencies necessary to effectively establish and execute all relevant responsibilities outlined above.

2. Senior officers shall be prepared to assess the need for formal debriefing in complex events, including knowledge of consultation resources, mutual aid agreements, and methods for access and activation.

3. Peer support personnel shall be prepared to function in both internal and external response modes, according to the protocols and agreements of the agency.

Accountability

1. Incident Commander (or senior officer of this agency when operating under mutual aid) shall hold responsibility for ensuring that all pertinent on-scene measures are proactively assigned and executed.

2. Incident Commander (or senior officer of this agency when operating under mutual aid) shall hold responsibility for ensuring that all pertinent post-incident assessments are made, and that appropriate measures are identified and executed.

3. Sector officers are accountable for elements of their responsibilities outlined in this section, and the execution of those elements of responsibility will be specifically considered in operational review of the incident.

4. Executive management shall be accountable for establishment and maintenance of appropriate mutual aid agreements, both for requests of service and for response to requests by other agencies.

CORNELIUS H. SCANNELL, MD, MPH
JOHN R. BALMES, MD

PULMONARY EFFECTS OF FIREFIGHTING

From the Division of Occupational and Environmental Medicine
Department of Medicine
University of California, San Francisco
San Francisco, California

Reprint requests to:
Cornelius H. Scannell, MD, MPH
Division of Occupational and Environmental Medicine
Department of Medicine
University of California, San Francisco
Box 0843
San Francisco, CA 94143-0843

Smoke from any fire contains gases and particulate matter that are irritating to the respiratory tract. These toxicants are the combustion and pyrolysis products of the natural and/or synthetic materials that are burning. The increasing use of synthetic materials in building and furniture construction, however, has led to a heightened concern about pulmonary effects from inhalation of the combustion products of these materials. In fact, exposure assessment studies indicate that firefighters are frequently exposed to significant concentrations of respiratory tract irritants, including sulfur dioxide, hydrogen chloride, phosgene, oxides of nitrogen, aldehydes, and particulates[4,13,15a,40] The water solubility of irritant gases determines their site of deposition[2]—the more water-soluble gases (e.g., hydrogen chloride) have a greater deposition on the moist surfaces of the upper airways and are less likely to penetrate in high concentration to the distal lung. Relatively insoluble gases (e.g., phosgene) are more likely to reach the alveoli. Thus, inhalation of water-soluble gases tends to cause proximal airway injury (bronchitis), and inhalation of insoluble gases tends to cause lung injury (bronchiolitis and pneumonitis). Often, however, inhaled smoke contains both types of irritant gases. Chlorinated hydrocarbons, for example, give rise to both hydrogen chloride and phosgene when burned. Particulates suspended in smoke may cause airway or deep lung injury, depending on the size distribution of the inhaled aerosol. Finally, direct thermal injury to the airways is a rare but possible outcome of exposure to fires. The upper respiratory tract is an effective thermal sink, but when this protective mechanism is overcome, as may occur during the inhalation of steam, damage to the distal airways and even the lung parenchyma can occur.

Occupational exposure to smoke may produce a wide range of pulmonary effects in firefighters, ranging from acute changes in lung function to increased mortality from nonmalignant respiratory disease. Although improved respiratory protection has become available to protect firefighters from smoke inhalation, the inconvenience, weight, and decreased mobility resulting from the use of the typical self-contained breathing apparatus often make compliance less than adequate.

ACUTE EFFECTS

Acute pulmonary effects of firefighting have been relatively well-studied over the past two decades. Although multiple studies have documented increased respiratory symptoms and acute physiologic changes, several studies have demonstrated little or no effect, probably due to relatively lower exposure to irritants as a result of greater use of respiratory protective equipment in modern structural firefighting.

Given the potential for firefighter inhalation of systemic asphyxiants, such as carbon monoxide and hydrogen cyanide, in addition to the respiratory irritants described earlier, it is not surprising that transient hypoxemia has been reported as a result of smoke inhalation. Genovesi et al. documented moderately severe hypoxemia in 19 of 21 mostly asymptomatic firefighters who were exposed to dense smoke from a fire in which large quantities of polyvinylchloride (PVC)-containing materials were burned.[12] The values for PaO_2 obtained within 10 hours of exposure were ≤ 76 mm Hg for 14 of the firefighters. This hypoxemia was transient, with nearly complete reversal within 24 hours. Normal pulmonary function tests obtained 1 month later demonstrated that this transient hypoxemia was not related to previous underlying pulmonary disease.

Taskin et al. administered a standardized respiratory questionnaire and detailed pulmonary function tests to the cohort of 21 Los Angeles firefighters studied by Genovesi et al. 1 month following the fire involving exposure to the combustion products of PVC.[38] The pulmonary function tests obtained included spirometry, airways resistance, lung volumes, single-breath nitrogen washout, and closing volume. The results of these studies were compared with those obtained in a sample of nonfirefighters residing in the Los Angeles area matched for anthropomorphic characteristics and smoking status. The frequency of respiratory symptoms and the results of the pulmonary function tests were similar in the firefighters and matched controls. The mean closing volume was actually higher in the controls. These findings suggested that although fighting fires may result in acute hypoxemia secondary to smoke inhalation (as was noted by Genovesi et al.), a single episode need not cause persistent respiratory symptoms or abnormalities of pulmonary function.

Although most episodes of firefighter smoke inhalation probably do not cause much lung injury, a single severe episode can lead to chronic respiratory impairment. Loke et al. obtained baseline maximum expiratory flow-volume curves in 54 firefighters while breathing (a) air and (b) a helium-oxygen mixture.[20] A ≤ 20% improvement in maximum expiratory flow at 50% of the vital capacity while breathing the helium-oxygen mixture is consistent with the presence of small airways obstruction. Evidence of small airways obstruction using this method was observed in 9 of 26 smoking firefighters and 2 of 15 nonsmoking firefighters. Both of the nonsmoking firefighters with small airways obstruction had been firefighting for ≥ 25 years, but neither had respiratory complaints. In addition to baseline studies, the authors retested 7 firefighters immediately after a mild exposure to smoke from a building fire, and no significant acute changes in pulmonary function were noted in this group. However, 1 firefighter, who was trapped in a building and thereby sustained heavy exposure to smoke, developed a severe obstructive ventilatory impairment that was still present more than 2 years after the fire. The severity of this firefighter's smoke inhalation was such that he required hospitalization for acute care, his admission carboxyhemoglobin level was 42%, a transbronchial lung biopsy later revealed coagulation necrosis, and he eventually developed radiographic evidence of saccular bronchiectasis.

Unger et al. studied a cohort of 30 firefighters with spirometry immediately after a severe smoke exposure and then 6 weeks and 18 months later.[41] The smoke exposure was the result of a fire that consumed a chemical warehouse containing chlorinated hydrocarbons, including pesticides. All firefighters studied were evaluated at a local emergency department. The spirometric results were compared with predicted values and with those from a group of closely matched control subjects. Surprisingly, no statistically significant differences between the acute postexposure spirometry values and those recorded at 6 weeks and 18 months were found. However, a trend toward an increased rate of volume loss in the forced vital capacity (FVC) and forced expiratory volume in 1 second (FEV_1) was noted. Furthermore, a significantly lower mean FVC was observed in the exposed firefighters compared with the mean predicted value, and both mean FEV_1 and FVC values were reduced compared with the control subjects.

Musk et al. were the first investigators to determine the extent to which acute changes in pulmonary function occurred during routine firefighting and to relate these changes to symptoms and indices of smoke exposure.[27] Spirometry and expired carbon monoxide concentrations were measured in 39 Boston firefighters at the beginning of each workshift and upon return of the firefighters to their quarters following any incident. Each subject served as his own control. Samples of the smoke encountered in the fires were collected by a personal monitoring apparatus mounted in a firefighter's uniform. The collected samples were analyzed for HCl, HCN, NO_2, CO, and total suspended particulate matter. Information covering 137 fire attendances was collected. The magnitude of the decline in FEV_1 was related to the severity of smoke exposure as estimated by the firefighters, the presence of irritative symptoms, and the concentration of total suspended particulate matter. Measured exhaled CO was also correlated with reported severity of smoke exposure. Decreases in FEV_1 in excess of 100 ml were recorded in 30% of observations. Changes in FEV_1 resulting from a second exposure to smoke on the same tour of duty were greater when smoke exposure at the previous fire was heavy. The largest postexposure decline in FEV_1 (1.4 L) occurred after a second fire in which the firefighter was exposed to burning plastics. Insufficient data were available to establish whether levels of HCl, HCN, or NO_2 were associated with change in FEV_1. This innovative study clearly documented the potential for routine firefighting to cause acute respiratory effects. The investigators noted that at the time of their study, a self-contained breathing apparatus would be worn by Boston firefighters only when smoke conditions were perceived to be dangerous. At least in part due to the results of this study, the Boston fire department began to require the use of respiratory protective equipment with all smoke exposures.

Since the study of Musk et al., several other reports have been published about the acute effects of routine firefighting on pulmonary function. Sheppard et al. studied 29 firefighters from a single San Francisco fire station.[32] In addition to obtaining pre-shift, post-shift, and post-fire spirometry, these investigators measured baseline nonspecific airway responsiveness by methacholine inhalation challenge and attempted to repeat methacholine challenge testing within 36 hours after every fire in which FEV_1 or FVC fell by $\geq 10\%$. Eighteen of 76 measurements obtained within 2 hours after a fire (24%) showed a > 2 standard deviation (SD) fall in FEV_1 and/or FVC compared to 2 of 199 obtained after routine workshifts without fires. On 13 of 18 occasions when across-shift spirometry decreased significantly, repeat spirometry was obtained 3 to 18.5 hours after fires; on four of these occasions FEV_1 and/or FVC were still > 2 SD below baseline. The occurrence of post-fire decrements in spirometry was not predicted by the baseline level of airway responsiveness. Due to logistical difficulties, repeat methacholine challenge was obtained in only two firefighters with > 10% post-fire declines in FEV_1 and FVC. In these two subjects, airway responsiveness was substantially increased. The investigators concluded that routine firefighting is associated with a high incidence of acute decrements in lung function. The persistence of the decrements for many hours after smoke exposure, the lack of association with baseline airway responsiveness, and the association with post-fire increases in airway responsiveness in two subjects suggested to the authors that these decrements in lung function were not due merely to irritant bronchoconstriction.

Sherman et al. were able to replicate the Sheppard et al. study design in a small group of Seattle firefighters and were more successful in obtaining post-fire repeat methacholine challenge testing.[34] Spirometry and methacholine challenge testing were performed on 18 active Seattle firefighters before and 5 to 24 hours after firefighting. Fire exposure was defined as the total number of hours spent without a self-contained breathing apparatus at the firesite. After firefighting, the mean percent predicted values for FEV_1 and forced expiratory flow at 25% to 75% FVC (FEF_{25-75}) predicted significantly decreased by means of 3.4% and 5.6%, respectively. The mean level of nonspecific airway responsiveness increased significantly after firefighting. This increase in airway responsiveness persisted after adjustment for prechallenge level of lung function. Analysis of variance revealed that the change in airway responsiveness was greatest in those with moderate exposure and was not related to smoking or years of firefighting. Like Sheppard et al., these investigators concluded that routine firefighting is associated with acute decrements in spirometric function and an acute increase in airway responsiveness.

Large et al. also looked at the short-term effects of smoke exposure on spirometry in a group of 60 Pittsburgh firefighters.[19] All 60 firefighters underwent baseline spirometry after a minimum of 4 off-duty days, and 22 firefighters were retested immediately upon return to their station following exposure to a house fire. Small but significant post-exposure declines in FEV_1 and FEF_{25-75} were observed. Estimated duration of smoke exposure was not significantly associated with observed change in FEV_1, but the investigators reported that the intensity of smoke exposure was difficult to assess by means of their post-fire questionnaire due to varying self-contained breathing apparatus use. Because two firefighters experienced exaggerated decrements in spirometric function compared to the group as a whole, the investigators concluded that there may be a small subgroup of firefighters with increased susceptibility to smoke inhalation.

Brandt-Rauf et al. conducted another evaluation of the acute effects of firefighting on spirometry in a cohort of Buffalo firefighters.[5] The fire locations included residential structures, industrial buildings, and an automobile. There were no significant across-fire differences in FEV_1 and FVC for the entire group of 37 firefighters for whom paired pre-fire and post-fire values were available. When the subgroup of 14 firefighters who did not use any respiratory protective equipment at any time during the course of study was analyzed separately, however, significant across-fire declines in both FEV_1 and FVC were observed. Although the investigators also obtained data on the exposure of their subjects to various toxicants, they did not report whether associations between level of exposure and change in spirometric function were present.

MECHANISMS OF ACUTE EFFECTS

The mechanisms underlying the acute changes in ventilatory function and airway responsiveness that have been documented in firefighters are not fully understood. Reflex bronchoconstriction and/or decreased inspiratory capacity as a result of stimulation of irritant receptors is one mechanism by which inhaled toxicants can cause acute impairment of ventilatory function. Sulfur dioxide and cigarette smoke can induce reflex bronchoconstriction mediated at least in part by muscarinic neurons[33] and ozone can induce a more restrictive-type of ventilatory impairment, apparently mediated by C-fiber stimulation.[8] Smoke-induced injury/inflammation of the airway epithelium may also play a role. Airway inflammation can lead to decreased airway diameter due to bronchial mucosal edema. Although the mechanism(s) by which inhaled irritants induce acute increases in airway responsiveness is also not entirely clear,[2] airway narrowing can shift the dose-response curves for methacholine or histamine challenges to the left, increased epithelial permeability[23] may lead to enhanced exposure of sensory neuroreceptors, and epithelial injury may involve loss of epithelial-derived smooth muscle relaxing factors.

An interesting controlled human exposure study was conducted by Chia et al. in an effort to learn more about the mechanisms of smoke-induced increases in airway responsiveness.[7] Spirometry and histamine inhalation challenge tests were obtained in two groups of 10 firefighters 1 hour before and 1 hour after exposure to smoke in a chamber. The subjects in one group were regular firefighters (average of 4.6 years of firefighting), and the subjects in the other group were recruit firefighters (no firefighting experience). All subjects were cigarette smokers, but none had airway hyperresponsiveness prior to exposure. Eight (80%) of the regular firefighters, however, developed an increase in airway responsiveness after exposure in the smoke chamber. After 6 hours, 3 of the firefighters still had increased responsiveness, and at 24 hours, these firefighters had persistent decrements in ventilatory function. In contrast, none of the 10 recruits showed increased airways responsiveness after the exposure. Multiple linear regression analysis showed duration of service as a firefighter to be a significant contributing factor to the smoke-induced change in airway responsiveness. The investigators speculated that chronic smoke-induced airway injury/inflammation in firefighters may enhance their risk of acute smoke-induced increases in airway responsiveness.

A somewhat related study of smoke-induced airway hyperresponsiveness was conducted by Kinsella et al.[18] These investigators studied 13 fire victims (rather than firefighters) who required treatment after smoke inhalation. The fire victims underwent pulmonary function testing and histamine inhalation challenges within 3 days of smoke inhalation and then 3 months later. The initial testing showed most of the victims to have airway hyperresponsiveness, but 3 months later, 10 of the subjects showed decreased responsiveness (immediate post-exposure median PC_{20}, 1.65 mg/ml, and 3-month post-exposure median PC_{20}, 5.43 mg/ml). Specific airways conductance and FEV_1, however, did not improve significantly over the 3-month observation period, and several of the subjects still had evidence of airways obstruction. There was a strong correlation between an index of smoke exposure, carboxyhemoglobin concentration, and the initial values for specific airways conductance. The investigators concluded that airways obstruction after smoke inhalation may be more common and more persistent than in generally recognized. Like Chia et al., they also speculated that airway epithelial injury/inflammation was likely an important contributing factor in smoke-induced lung function changes.

Inhalation of irritant chemicals has clearly been documented to cause airway epithelial injury and inflammation, and a number of case reports describe emergency responders, including firefighters, who have developed irritant-induced asthma following short-term, high-intensity exposure to such chemicals. One of the few epidemiologic studies of such an exposure was reported by Markowitz et al.[21] These investigators conducted a retrospective cohort study of 80 firefighters exposed to burning PVC as well as 15 nonexposed firefighter control subjects. Administering a questionnaire 5 to 6 weeks after the exposure, the investigators found a significant increase in respiratory symptoms (chest pain, wheezing, coughing, and shortness of breath) in the exposed cohort compared to the control subjects. There was some evidence of an exposure-response relationship among the exposed firefighters. Because the subjects involved in the fire also reported mucosal irritative symptoms consistent with HCl exposure, a compound likely generated in high concentration, the investigators concluded that this agent may have been responsible for much of the reported lower respiratory tract symptomatology. The results of this study provide evidence that persistent respiratory symptoms can occur after sufficient exposure to a specific material or mixture that is capable of causing chemical injury to the airway epithelium.

CHRONIC EFFECTS

The issue of whether recurrent exposure to smoke from routine firefighting in the modern era leads to chronic pulmonary effects has not been definitively resolved. Although earlier studies provided evidence supporting increased risks for both chronic respiratory symptoms and deficits in pulmonary function, more recent studies involving populations characterized by a greater use of respiratory protective gear have tended not to demonstrate chronic effects of smoke exposure.

Markowitz reported the results of the long-term follow-up of the cohort of firefighters exposed to burning PVC (see description in preceding section).[22] Twenty-two months after the fire, the initially studied subjects were asked to complete a follow-up questionnaire. A total of 55 exposed firefighters were studied at both time points. Exposed subjects reported significantly more frequent and severe respiratory symptoms at both time points than did firefighter controls. In longitudinal analyses, a number of symptoms persisted over time, and initially reported acute symptom scores were significantly correlated with chronic symptom scores. At 22 months after the PVC fire, approximately 18% of exposed firefighters, compared with none of the controls, reported that during the interval time a physician had told them that they had either asthma or bronchitis.

Horsfield et al. followed respiratory symptoms in a group of 96 British firefighters (31 nonsmokers, 40 smokers, and 25 former smokers) over 4 years.[16] A control group of 69 male nonsmokers from a variety of occupations was also followed over this period. A history of symptoms and of smoking habits was obtained on entry to the study, then every 6 months for 2 years, and then annually for a further 2 years. All those remaining in the study after 4 years were interviewed and history of use of a self-contained breathing apparatus and of symptomatic smoke inhalation were obtained. Symptom frequency was least in control subjects (all nonsmokers), intermediate in firefighters who were never or former smokers, and greatest in smoking firefighters. On entry to the study, respiratory symptoms were 3.9 times greater in currently smoking firefighters and 2.3 times greater in those with past symptomatic smoke inhalation, as compared to nonsmokers who had never experienced symptomatic smoke inhalation. In current smokers with a history of symptomatic smoke inhalation, symptoms were increased 9.1 times, suggesting a multiplicative effect. During the 4-year study period, symptom frequency was increased 2.9 times in nonsmoking firefighters compared to controls and 1.8 times in smoking firefighters compared to the nonsmoking coworkers. In firefighters who are nonsmokers and who had not experienced symptomatic smoke inhalation, symptom frequency was similar to that observed in the control subjects. The investigators concluded that occupational smoke inhalation may cause long-term symptoms in firefighters and that the use of a self-contained breathing apparatus may prevent the development of such symptoms.

Sidor and Peters assessed the prevalence of chronic nonspecific respiratory disease in a cross-sectional study of 1768 Boston firefighters.[35] Spirometry and a standardized respiratory symptom questionnaire were used to classify firefighters into one of two categories, no disease or chronic nonspecific respiratory disease (including those with chronic nonproductive cough, chronic bronchitis, asthma, and chronic obstructive lung disease). Relevant exposure information was also obtained, such as number of times oxygen was used at fires, number of episodes of symptomatic smoke inhalation, number of hospitalizations for smoke inhalation. Experienced firefighters had a higher rate of chronic nonspecific respiratory disease than new firefighters of the same age group, although cigarette smoking partially obscured the occupational effect. Several indicators of occupational exposure, including number of episodes of symptomatic smoke inhalation, were associated with higher chronic nonspecific respiratory disease rates. Significantly higher disease rates were found in firefighters who smoked or had smoked cigarettes in the past. Although the investigators found evidence of a healthy worker effect (i.e., lower prevalence of chronic respiratory disease in firefighters than in the general population), they also demonstrated that firefighters incur increased risk of respiratory disease due to their occupation, especially if they also smoke cigarettes.

In a companion paper, Sidor and Peters reported their analysis of spirometric data for the cohort of 1786 Boston firefighters.[36] The majority of firefighters experienced increased mucus production and a general malaise after exposure to smoke at fires. Prolonged mucus secretion after occupational exposures was associated with lower ventilatory function. Cigarette smoking was also strongly associated with mucus production and decreased ventilatory function. Although years of employment in the fire department was associated with a lower FEV_1, it was actually years of nonfirefighting duty that was responsible for this effect. In fact, smoke exposure variables such as number of "pastings" or "shellackings" (i.e., episodes of symptomatic smoke inhalation) tended to predict larger ventilatory capacity among the firefighters. The investigators speculated that this apparent paradoxical finding was due to self-selection by firefighters; only healthy firefighters can fight fires. Firefighters with respiratory impairment seek less active duty and migrate to support functions. As a result, cross-sectional data are likely to underestimate the effects of firefighting on lung function.

Peters et al. also reported the results of repeat questionnaire administration and spirometry in their Boston firefighter cohort after 1 year of follow-up.[29] Repeat data were available for 1430 of the 1768 firefighters initially tested in 1970–71. The annual rate of decline in ventilatory function observed for the study population was more than twice the expected rate (68 ml for FEV_1 and 77 ml for FVC) and was significantly related to frequency of fire exposure. Again, the number of episodes of symptomatic smoke inhalation during the period of observation was not significantly associated with the rate of decline in FEV_1 or FVC. The 64 firefighters who required hospitalization for smoke inhalation, however, did have a greater annual rate of decline in ventilatory function than the group as a whole (98 ml for FEV_1 and 90 ml for FVC). The investigators interpreted the results of this study as strongly suggesting that occupational exposures contribute to accelerated loss of lung function in firefighters.

Longitudinal follow-up of the Boston firefighter cohort was continued over a 6-year period. The results of the 3-year follow-up were reported by Musk et al.[24] Repeat spirometry data were available in 1974 for 1,146 of the original 1,768 subjects. In contrast to the accelerated rate of decline in ventilatory function related to frequency of fire exposure observed over the first year of the longitudinal study, the investigators observed annual rates of decline over the 3-year follow-up period (30 ml for FEV_1 and 40 ml for FVC) that are within the expected range for the general population. Moreover, the rates of decline in FEV_1 and FVC were not related to the number of fires fought or to other indices of occupational smoke exposure. The investigators felt that the inconsistency in their 1-year and 3-year follow-up results was in part due to selection factors within the fire department. They noted that through promotions or job changes within the department, firefighters with respiratory symptoms and impaired ventilatory function tended to be selectively removed from active firefighting duty prior to retirement. In fact, the subjects who were transferred to nonfirefighting duty did have a greater annual rate of decline in ventilatory function than the group as a whole. Further, the subjects lost to follow-up at 3 years were older and had a slightly greater rate of decline after 1 year than the subjects who participated in the 3-year follow-up study. The investigators also noted that there had been an increased use of respiratory protective equipment by firefighters over the latter 2 years of the observation period.

In a companion study of retirees, Musk et al. attempted to assess further whether a healthy worker survival bias was influencing the longitudinal results from the Boston firefighter cohort.[25] They administered questionnaires and spirometry to 109 firefighters who had retired between 1970 and 1975. The mean values for both FEV_1 and FVC in the retirees were only slightly lower than predicted (05% and 99%, respectively). The investigators again presented evidence that firefighters with respiratory symptoms and decreased ventilatory function were selectively removed from active firefighting duty prior to retirement. The subjects who were not actively fighting fires at the time of retirement had a greater proportion of former smokers, a greater prevalence of chronic bronchitis, and lower ventilatory function than those who were actively fighting fires. Subjects who retired with a shorter length of service were younger, had a greater prevalence of chronic bronchitis, and had lower ventilatory function than the other retirees. The investigators concluded that a healthy worker effect was indeed likely to be affecting the relationship between occupational exposure to fire smoke and rate of decline in ventilatory function in their study population.

In the final paper from the Boston firefighter study, Musk et al. reported the results of the 6-year follow-up.[28] The investigators were able to obtain follow-up data on 951 of the subjects who participated in the initial testing in 1970. During the 6 years of follow-up, the mean annual decrements in FEV_1 and FVC were 36 and 29 ml, respectively. At the end of the study in 1976, the mean values for both FEV_1 and FVC for the group as a whole were 98% of the predicted values. As expected, current cigarette smoking was associated with an increased prevalence of chronic bronchitis, reduced FEV_1 and FVC, and greater longitudinal declines in FEV_1 and FVC. Neither the 1976 values for FEV_1 and FVC nor the longitudinal declines in these parameters were correlated with any index of firefighting exposure in active firefighters. The investigators again concluded that these findings were due to the increased use of respiratory protective equipment and selection effects within the Boston Fire Department.

In contrast to the overall impression from the longitudinal study of the Boston Fire Department of a lack of effect of firefighting on annual decline in ventilatory function, another longitudinal study of Boston-area firefighters by Sparrow et al. provided evidence of a chronic effect of firefighting on lung function.[37] This investigation involved 168 firefighters and 1474 nonfirefighters enrolled in the Veterans' Administration (VA) Normative Aging Study. A questionnaire was administered and spirometry was obtained at intake and 5 years later. Firefighters had a significantly greater decline in both FEV_1 and FVC than nonfirefighters. This occupational effect could not be explained by differences in age, height, smoking status, or initial level of lung function between the two groups. Overall, firefighters reported more respiratory symptoms and disease in the follow-up period than did nonfirefighters, but the number of subjects developing respiratory symptoms or disease were few, and no significant differences between the groups were demonstrated after adjustment for age and smoking status. It is important to note that the firefighters participating in the VA Normative Aging Study were initially recruited during 1963–68 and retested during 1968–73. Thus, the occupational exposures of the participating firefighters may have been greater than those of many of the subjects in the Boston Fire Department longitudinal study conducted between 1970 and 1976.

A lack of effect of modern firefighting on annual decline in ventilatory function was supported by two more recent British studies. Douglas et al. conducted a prospective study of ventilatory function in a sample of London firefighters.[10] These investigators interviewed and obtained spirometry in 1006 firefighters in 1976; 895 were tested a second time 1 year later. The mean values for both FEV_1 and FVC were greater than the mean predicted values at both time points. The mean declines in FEV_1 and FVC over the 1-year follow-up period were 92 ml and 107 ml, respectively. Multiple regression analysis indicated that FEV_1 and FVC fell more rapidly in those aged over 40 years and that cigarette smoking had the expected strong negative effect on these measures of ventilatory function. No association was demonstrated between indices of occupational exposure to fire smoke and either respiratory symptoms or ventilatory function. Only among men with over 20 years of service was there possibly any evidence (not statistically significant) of an effect from duration of employment.

In addition to the respiratory symptom paper reviewed earlier, Horsfield et al. followed the lung function of the group of 96 West Sussex firefighters for 1 to 4 years.[17] Relatively sophisticated pulmonary function testing (spirometry with expiratory flow-volume curves, determination of lung volumes by body plethysmography, and nitrogen washout) was repeated every 6 months for the first 2 years and annually for the last 2 years. The results of pulmonary function testing were expressed in terms of the rate of change with time. As expected, many of the parameters deteriorated in both firefighters and controls, but the rate of deterioration was greater in the controls than in the firefighters for FEV_1, FVC, peak expiratory flow, flows at 50% and 25% of vital capacity, airways resistance, and the ratio of residual volume to total lung capacity. Alveolar mixing efficiency (AME), a measure of small airways function, did not change significantly over the study period in any group. Nonsmoking firefighters had the highest mean value of AME, decreasing through exsmoking firefighters, controls, and smoking firefighters. The investigators concluded that their results showed no evidence of chronic lung damage due to modern firefighting and that firefighters as a group may have a lower rate of decline of lung function with age than the general population. They attributed these findings to the selection of fit men for firefighting service, continued physical training of active-duty firefighters, and the regular use of respiratory protective equipment.

Because urban structural firefighting in the modern era is characterized by widespread use of respiratory protective equipment (usually a self-contained breathing apparatus), studies of wildland firefighters, who do not generally wear such equipment, are relevant to any consideration of the chronic effects of occupational smoke exposure on lung function. The chapter by Rothman and Harrison (page 857) describes current knowledge about the respiratory health effects of wildland firefighting.

MORTALITY

Firefighters are occupationally exposed on a chronic basis to combustion products that are both carcinogenic and irritating to the respiratory tract. It is reasonable to suspect, therefore, that they may be at increased risk of death due to lung cancer and nonmalignant respiratory disease. Despite the studies described earlier which show both acute and chronic pulmonary effects of smoke inhalation, excess mortality due to respiratory disease has not been consistently demonstrated among firefighters as compared to the general population. A potential problem with mortality studies of firefighters is the healthy worker effect. One strategy to avoid this problem is to use police as the comparison population, because police officers may be similar to firefighters in terms of the physical fitness requirements for employment. Studies that have used this strategy have generally found that firefighters have higher mortality due to nonmalignant respiratory disease than do police.

One of the first mortality studies to apply modern epidemiologic techniques was that of Musk et al., which described the mortality experience of all firefighters employed for more than 3 years by the Boston Fire Department between 1915 and 1975 (n = 5,655).[26] There were 2,470 total deaths in the cohort at the end of the follow-up period. The observed death rates due to various causes were compared to those expected in the male population of Massachusetts and of the entire United States. The all-cause standardized mortality ratio (SMR) was 0.91. The SMRs for nonmalignant respiratory deaths, cardiovascular deaths, and cancer deaths were 0.93, 0.86, and 0.83, respectively. In contrast, the SMR for deaths due to accidents was 1.35. The authors speculated that the strict job entry criteria for firefighters, including pre-employment physical examinations, led to the screening out of less healthy individuals from the profession.

Vena et al. studied the mortality experience of 1,867 white male firefighters who were employed by the City of Buffalo for a minimum of 5 years with at least 1 year as a firefighter.[42] Vital status was determined for 99% of the cohort, resulting in 470 observed deaths. The firefighter mortality experience was again characteristic of a healthy worker population. When compared to the general U.S. population, all-cause mortality was lower than expected (SMR = 0.95), and mortality due to all nonmalignant respiratory diseases was significantly lower than expected (SMR = 0.48). Morality from respiratory cancers was also lower than expected (SMR = 0.94). Beaumont et al. performed a similar analysis of the mortality experience of 3066 San Francisco firefighters employed between 1940 and 1970 using U.S. death rates for comparison.[3] These investigators again found that deaths due to nonmalignant respiratory diseases occurred significantly less often than expected (SMR = 0.63) and that deaths due to lung cancer were also lower than expected (SMR = 0.84). Subsequent studies which used the general population for comparison of death rates have similarly found that firefighters do not have excessive mortality due to either nonmalignant respiratory disease or lung cancer[1, 14, 15, 39] As noted earlier, all of these studies suffer from possible confounding due to the healthy worker effect.

When police have been used as a reference population, evidence that firefighters are at increased risk of respiratory disease has been found. The first such study was conducted by Feuer and Rosenman.[11] These investigators compared the proportionate mortality of all New Jersey firefighters participating in the state retirement system between 1974 and 1980 to that of three reference populations: the general U.S. population, the New Jersey general population, and New Jersey police. There were 901 firefighter deaths during the study period. The firefighters had a significant increase in nonmalignant respiratory disease only when the police were used as a reference group (proportionate mortality rate = 1.98), supporting the concept that the healthy worker effect must be considered when evaluating firefighter mortality.

The results of this initial firefighter–police comparison were confirmed by the work of Rosenstock et al. who conducted a mortality analysis of firefighters in three northwestern U.S. cities using the mortality experience of both the general U.S. population and a cohort of police officers for comparison.[30] The firefighters were employed between 1945 and 1980 and experienced 886 deaths through the end of 1983. When compared with the U.S. population, the firefighters had a significantly reduced risk of dying from all causes and cardiovascular diseases and had a nonsignificantly reduced risk of death due to nonmalignant respiratory diseases (SMR = 0.88). Compared with police, however, the firefighters had a significant excess of deaths from nonmalignant respiratory diseases (SMR = 1.41). Longer follow-up of this cohort through 1989 showed persistence of an increased risk of death due to nonmalignant respiratory disease in comparison to police, but this difference was no longer significant.[9] The authors speculated that the reduced effect of firefighting on risk of nonmalignant respiratory disease observed in their later study was due, in part, to increased use of respiratory protection since the 1970s. Taken together, the results of these firefighter-police comparison studies suggest that firefighters are probably at increased risk of dying from nonmalignant respiratory disease because of occupational factors. This increased risk may have been missed in other studies because the use of a general reference population does not control for the healthy worker effect.

Studies using police as the comparison population have failed to find an increased risk for lung cancer among firefighters. The SMR for lung cancer, with police as the reference population, was 1.02 in the New Jersey study and 0.86 in the initial report of the northwestern U.S. firefighter study. In addition, a study of cancer incidence among Massachusetts firefighters from 1982 to 1986 was conducted by Sama et al., using state cancer registry data.[31] For the 315 male cancer cases among firefighters reported to the registry during the study period, the standardized morbidity odds ratio was elevated for lung cancer, using both the state general population and police as reference groups (1.22 and 1.33, respectively), but neither elevation was statistically significant.

One additional study of firefighter mortality that provides valuable information is that of Burnett et al.[6] These investigators analyzed data from the National Occupational Mortality Surveillance system that collects death certificate information, including occupation, from 27 states. A total of 5744 white male firefighter deaths were reported between 1984 and 1990. Proportionate mortality ratios for lung cancer and chronic obstructive pulmonary disease were 1.02 and 0.83, respectively. This study is notable for its size and wide geographic coverage. The lack of evidence for increased mortality due to respiratory disease among firefighters is reassuring.

SUMMARY

Smoke from any fire contains gases and particulate matter irritating to the respiratory tract. Occupational exposure to smoke may produce both acute and chronic pulmonary effects in firefighters. Although acute respiratory symptoms and lung function changes have been well-documented in firefighters after smoke exposure, several recent studies have demonstrated little or no effect, presumably because of increased use of respiratory protective equipment in modern structural firefighting. Similarly, although earlier studies of the chronic pulmonary effects of firefighting provided evidence supporting increased risks for both respiratory symptoms and deficits in lung function, more recent studies have not. When compared to the general population, firefighters tend to have decreased overall mortality, probably as a result of the healthy worker effect. However, when compared to another occupational group with comparable fitness requirements, police, firefighters have been shown to have an increased risk of death due to nonmalignant respiratory disease in several studies. Despite exposure to respiratory tract carcinogens in fire smoke, no increased risk of lung cancer has been associated with firefighting.

REFERENCES

1. Aronson KJ, Tomlinson GA, Smith L: Mortality among firefighters in metropolitan Toronto. Am J Ind Med 26:89–101, 1994.
2. Balmes J: Acute pulmonary injury from hazardous materials. In Sullivan JB, Krieger GR (eds): Hazardous Materials Toxicology. Baltimore, Williams & Wilkins, 1992, pp 425–432.
3. Beaumont JJ, Chu GST, Jones JR, et al: An epidemiologic study of cancer and other causes of mortality in San Francisco firefighters. Am J Ind Med 19:357–372, 1991.
4. Brandt-Rauf PW, Fallon LF, Tarantini T, et al: Health hazards of firefighters: Exposure assessment. Br J Ind Med 45:606–612, 1988.
5. Brandt-Rauf PW, Cosman B, Fallon LF, et al: Health hazards of firefighters: Acute pulmonary effects after toxic exposures. Br J Ind Med 46:209–211, 1989.

6. Burnett CA, Halperin WE, Lalich NR, Sestito JP: Mortality among fire fighters: A 27 state survey. Am J Ind Med 26:831–833, 1994.
7. Chia KS, Jeyaratnam J, Chan TB, Lim TK: Airway responsiveness of firefighters after smoke exposure. Br J Ind Med 47:524–527, 1990.
8. Coleridge JCG, Coleridge HM, Schelegle ES, Green JF: Acute inhalation of ozone stimulates bronchial C-fibers and rapidly adapting receptors in dogs. J Appl Physiol 74:2345–2352, 1993.
9. Demers PA, Heyer NJ, Rosenstock L: Mortality among firefighters from three northwestern United States cities. Br J Ind Med 49:664–670, 1992.
10. Douglas DB, Douglas RB, Oakes D, Scott G: Pulmonary function of London firemen. Br J Ind Med 42:55–58, 1985.
11. Feuer E, Rosenman K: Mortality in police and firefighters in New Jersey. Am J Ind Med 9:517–527, 1986.
12. Genovesi MG, Taskin DP, Chopra S, et al: Transient hypoxemia in firemen following inhalation of smoke. Chest 71:441–444, 1977.
13. Gold A, Burgess WA, Clougherty EV: Exposure of firefighters to toxic air contaminants. Am Ind Hyg Assoc J 39:534–539, 1978.
14. Grimes G, Hirsch D, Borgeson D: Risk of death among Honolulu firefighters. Hawaii Med J 50(3):82–85, 1991.
15. Guidotti TL: Mortality of urban firefighters in Alberta, 1927–1987. Am J Ind Med 23:921–940, 1993.
15a. Haponik EF: Clinical smoke inhalation injury: Pulmonary Effects. Occup Med State Art Rev 8:431–468, 1993.
16. Horsfield K, Cooper FM, Buckman PP, et al: Respiratory symptoms in West Sussex firemen. Br J Ind Med 45:251–255, 1988.
17. Horsfield K, Guyatt AR, Buckman MP, Cumming G: Lung function in West Sussex firemen: A four year study. Br J Ind Med 45:116–121, 1988.
18. Kinsella J, Carter R, Reid WH, et al: Increased airways reactivity after smoke inhalation. Lancet 337:595–597, 1991.
19. Large AA, Owens GR, Hoffman LA: The short-term effects of smoke exposure on the pulmonary function of firefighters. Chest 97:806–809, 1990.
20. Loke J, Farmer W, Matthay RA, et al: Acute and chronic effects of firefighting on pulmonary function. Chest 77:369–373, 1980.
21. Markowitz JS, Gutterman EM, Schwartz S, et al: Acute health effects among firefighters exposed to a polyvinyl chloride (PVC) fire. Am J Epidemiol 129:1023–1031, 1989.
22. Markowitz JS: Self-reported short- and long-term respiratory effects aong PVC-exposed firefighters. Arch Environ Health 44:30–33, 1989.
23. Minty BD, Royston D, Jones JG, et al: Changes in permeability of the alveolar-capillary barrier in firefighters. Br J Ind Med 42:631–634, 1985.
24. Musk AW, Peters JM, Wegman DH: Lung function in fire fighters: I. A three year follow-up of active subjects. Am J Public Health 67:626–629, 1977.
25. Musk AW, Peters JM, Wegman DH: Lung function in fire fighters: II. A five year follow-up of retirees. Am J Public Health 67:630–633, 1977.
26. Musk AW, Monson RR, Peters JM, Peters RK: Mortality among Boston firefighters, 1915–1975. Br J Ind Med 35:104–108, 1978.
27. Musk AW, Smith TJ, Peters JM, McLaughlin E: Pulmonary function in firefighters; Acute changes in ventilatory capacity and their correlates. Br J Ind Med 36:29–34, 1979.
28. Musk AW, Peters JM, Bernstein L, et al: Pulmonary function in firefighters: A six-year follow-up in the Boston fire department. Am J Ind Med 3:3–9, 1982.
29. Peters JM, Theriault GP, Fine LJ, Wegman DH: Chronic effect of fire fighting on pulmonary function. N Engl J Med 291:1320–1322, 1974.
30. Rosenstock I, Demers PA, Barnhart S: Respiratory mortality among firefighters. Br J Ind Med 47:462–464, 1990.
31. Sama SR, Martin TR, Davis LK, Kriebel D: Cancer incidence among Massachusets firefighters, 1982–1986. Am J Ind Med 18:47–54, 1990.
32. Sheppard D, Distefano S, Morse L, Becker C: Acute effects of routine firefighting and lung function. Am J Ind Med 9:333–340, 1986.
33. Sheppard D: Mechanisms of airway responses to inhaled sulfur dioxide. In Loke J (ed): Pathophysiology and Treatment of Inhalation Injuries. New York, Marcel Dekker, 1988, pp 49–65.
34. Sherman CB, Barnhart S, Miller MF, et al: Firefighting acutely increases airway responsiveness. Am Rev Respir Dis 140:185–190, 1989.
35. Sidor R, Peters JM: Prevalence rates of chornic non-specific respiratory disease in fire fighters. Am Rev Respir Dis 109:255–261, 1974.
36. Sidor R, Peters JM: Fire fighting and pulmonary function: An epidemiologic study. Am Rev Respir Dis 109:249–254, 1974.

37. Sparrow D, Bosse R, Rosner B, Weiss ST: The effect of occupational exposure on pulmonary function. Am Rev Respir Dis 125:319–322, 1982.
38. Taskin DP, Genovesi MG, Chopra S, et al: Respiratory status of Los Angeles firemen: One-month follow-up after inhalation of dense smoke. Chest 71:445–448, 1977.
39. Tornling G, Gustavsson P, Hogstedt C: Mortality and cancer incidence in Stockholm firefighters. Am J Ind Med 25:219–228, 1994.
40. Treitman RD, Burgess WA, Gold A: Air contaminants encountered by firefighters. Am Ind Hyg Assoc J 41:796–799, 1980.
41. Unger KM, Snow RM, Mestas JM, Miller WC: Smoke inhalation in firemen. Thorax 35:838–842, 1980.
42. Vena JE, Fiedler RC: Mortality of a municipal-worker cohort: IV. Firefighters. Am J Ind Med 11:671–684, 1987.

ANNE L. GOLDEN, PhD
STEVEN B. MARKOWITZ, MD
PHILIP J. LANDRIGAN, MD, MSc

THE RISK OF CANCER IN FIREFIGHTERS

From the Division of
 Environmental and Occupational
 Medicine
Department of Community
 Medicine
Mount Sinai School of Medicine
New York, New York

Reprint requests to:
Anne L. Golden, PhD
Assistant Professor
Box 1057
Mount Sinai School of Medicine
One Gustave L. Levy Place
New York, NY 10029

Cancer among firefighters has been an area of intensive investigation in occupational medicine for the past two decades. This research has been prompted by the recognition that firefighters are exposed in their work to high doses of multiple chemical carcinogens. The full extent of the occupational cancer risk of firefighters is not yet known. It is likely that in the years ahead, additional cancers will be found to be associated with exposures encountered by firefighters and that additional chemicals to which firefighters are already known to be exposed will be found to be carcinogenic. Despite the gaps in scientific knowledge, concern about excess cancer risk has resulted in the provision of disability benefits to firefighters under presumptive occupational cancer legislation in 15 states (Alabama, California, Illinois, Louisiana, Maryland, Massachusetts, Minnesota, Nevada, New Hampshire, North Dakota, Oklahoma, Rhode Island, Tennessee, Texas and Virginia) and in the city of New York.

A substantial body of literature now exists on the carcinogenic hazards of firefighting. Of particular concern are cancers that can be plausibly linked with specific toxic and carcinogenic chemical exposures to which firefighters are exposed in the course of their work: leukemia, lymphoma, multiple myeloma, melanoma, and cancers of the respiratory system, digestive system, genitourinary tract and brain.[30,36,45,64]

CARCINOGENIC EXPOSURES OF FIREFIGHTERS

Firefighters are routinely exposed to complex and dynamic mixtures of chemical substances that are contained in fire smoke and building debris.[14] Despite the large numbers of people employed in this occupation, the nature of these exposures is not well defined. Studies that have been completed to date, however, clearly demonstrate the presence of recognized and suspected human carcinogens in the breathing environment of firefighters at the fire scene.

The relative paucity of information about the exposures of firefighters is not surprising given the complexity of such exposures and the methods by which they are studied. Fires vary greatly in the nature of the materials burned, temperature, size, and ambient weather conditions.[14] The nature and concentrations of airborne exposures change at the fire scene over short distances and upon the stage of the fire. The actual exposures received by firefighters further depend on their job tasks at the fire and the type and use of respiratory protection. Finally, measurement of airborne exposures at fires presents formidable technical challenges in sampling methods, equipment, and logistics.[37]

While studies of firefighters have emphasized the importance of exposures at the fire scene, exposures at the firehouse, where firefighters spend long hours, also may have an impact on their risk of cancer. Diesel exhaust from fire trucks, especially if their engines are run in closed houses without direct venting to outside air, may lead to high levels of diesel exhaust emission particulates that are probably carcinogenic.[24] Many fire companies are located in old buildings, where deteriorating asbestos-containing insulation material may produce harmful levels of exposure to resident firefighters.

The following sections summarize the available data regarding carcinogenic exposures in the work environment of firefighters.

Benzene

Benzene is firmly established as a human carcinogen.[36] Numerous studies have shown that benzene is a common airborne contaminant in fire smoke and occurs in concentrations that are considered deleterious in the context of chronic exposures.

Treitman, Burgess, and Gold studied ambient environmental levels of a number of air contaminants, including benzene, at more than 200 structural fires in Boston in the mid-1970s.[69] Benzene was detected in 181 of 197 (92%) samples taken at fire scenes by air sampling units placed on the chests of firefighters. Half of the samples showed benzene over 1 part per million (ppm), the current OSHA permissible exposure level. Approximately 5% of the samples were above 10 ppm benzene.[69]

Lowry and colleagues studied firefighters' exposure to benzene at nearly 100 structural fires in Dallas in the early 1980s.[41] They found benzene at the majority of the fires but did not provide information about the levels measured. They also detected the presence of at least 70 organic chemical species regardless of whether synthetic materials were a major part of the materials burned.

Brandt-Rauf et al.[11] used personal portable sampling devices to measure exposures of 51 firefighters at 14 fires in Buffalo in 1986. The tubes of the sampling devices were attached to the firefighters' turnout gear, thereby representing ambient air outside the mask. Benzene was second only to carbon monoxide as the most common chemical substance detected at the fires.[11] It was detected in 18 of 26 samples from 12 of 14 fires. When detectable, the concentration of benzene ranged from 8.3 to 250 ppm. In only one sample where benzene was detected was its concentration below 10 ppm. Even when the smoke's intensity was rated as low, benzene was usually present in concentrations ranging from 22 to 54 ppm. The authors noted that respiratory protection was only partially used or not used at all at the fires judged to be of low smoke intensity.[11]

Jankovic and colleagues at the National Institute for Occupational Safety and Health (NIOSH) studied benzene and other exposures at 22 fires in the late 1980s, including 6 training fires, 15 residential fires, and 1 automobile fire.[37] Samples were collected via probes placed inside and outside the masks of working firefighters. In addition, industrial hygienists used a variety of sampling devices at the fire scene. Samples were taken separately during the two phases of a fire: knockdown and overhaul.

Half of the samples taken during the knockdown phase of the fire showed benzene in concentrations of 1–22 ppm. Of the 29 organic substances analyzed qualitatively by gas chromatography/mass spectrometry, benzene was the most common compound detected and was the only substance present in all eight samples.

To measure the efficacy of respiratory protection, samples for benzene were taken inside and outside the mask.[37] Surprisingly, the levels of benzene inside the mask were as high as those taken outside the mask and ranged from nondetectable to 21 ppm. The authors attributed this equivalence in benzene concentrations inside and outside the mask to partial or nonuse of the mask at the fire, especially after the initial phase of fire knockdown. They further suggested that benzene may be present only during the latter part of knockdown.[37]

During the overhaul phase of the fire, when respiratory protection is frequently removed, benzene concentrations were low, i.e., less than 1 ppm.[37]

Asbestos

Asbestos is universally recognized as a human carcinogen and has caused an excess in risk of a variety of cancers in numerous occupations.[36,63] The extent to which a firefighter has potential exposure to asbestos at the fire scene is an interesting and largely unanswered question. Since the building destruction caused by fires and the building demolition actively performed by firefighters during overhaul are likely to dislodge respirable asbestos fibers, the likelihood that firefighters have exposure to asbestos is high. However, the extent of such exposure is uncertain given intermittent exposure and use of respiratory protection.

Markowitz and colleagues at Mount Sinai School of Medicine in New York performed a cross-sectional study of 212 firefighters who had begun employment in the New York City Fire Department at least 25 years previously.[43] All participants had worked principally in ladder companies and, thus, had engaged in overhaul operations frequently. In addition, all participants had worked in locations in New York City where exposure to asbestos-containing materials was considered to be most common: high-rise office buildings, warehouses and factories, and poor neighborhoods with high fire activity in the 1960s.

Twenty of the 152 (13%) firefighters without prior exposure to asbestos had pleural thickening and/or parenchymal opacities on chest x-ray that represented characteristic sequelae of prior asbestos exposure. All of the chest-ray abnormalities were mild in degree. Twenty-two of the 60 (37%) firefighters with a history of exposure to asbestos prior to becoming a firefighter showed such radiologic abnormalities. Prevalence of radiographic abnormalities did not increase with duration of employment as a firefighter or duration from onset of employment, but the study criteria for subject selection assured a narrow range in these categories.

The authors concluded that long-term firefighters in urban areas may have significant exposure to asbestos and are at risk for asbestos-related diseases.[43] Although the Mount Sinai study was restricted to pleural and parenchymal fibrosis as outcomes of interest, the results are relevant to the issue of the risk of cancer for firefighters. The finding of excess risk of lung and pleural fibrosis due to asbestos among firefighters indicated that significant asbestos exposure has occurred in this group. Since significant asbestos exposure confers excess risk for selected cancers, it is reasonable to expect that firefighters have an increased risk of various cancers as a result of their exposure to asbestos.

No environmental study of ambient levels of asbestos at fire scenes has been undertaken. Jankovic et al. collected airborne fibers on cellulose filters at the scene of structural fires and analyzed these with polarized light microscopy.[37] The limit of detection was 0.4 fibers/ml. Fiber counts were higher during the overhaul phase than the knockdown phase of the fire. No asbestos fibers were detected, but cellulose and glass fibers were obtained. The investigators did not ascertain whether insulation materials were involved in any of the fires. They concluded that their results "do demonstrate the potential for exposures during overhaul when building materials contain asbestos."[37]

Polycyclic Aromatic Hydrocarbons

Polycyclic aromatic hydrocarbons (PAHs) are a class of organic substances that have been implicated as the carcinogenic substances in coal tar pitches, coal tar, and selected mineral oils.[36] They have been associated with excess risk of a variety of cancers, including cancer of the skin, lung, kidney, and bladder.[36]

Given the combustion of diverse materials at fires, it is likely a priori that firefighters would be exposed to significant levels of PAHs. Earlier studies of airborne contaminants at fires concentrated on the measurement of acute irritants and asphyxiants, ignoring the presence of PAHs. In their recent study, Jankovic et al. evaluated the presence of PAHs at the scene of fires.[37] All 14 PAHs measured, including benz(a)pyrene, were present at mean values of 3–63 $\mu g/m^3$ during the knockdown phase of the fire. Concentrations of PAHs during overhaul were considerably lower than during knockdown and were similar to those seen in ambient air in the absence of fire.

Formaldehyde

Formaldehyde is considered a probable human carcinogen.[36] In animal experiments, formaldehyde has caused cancer of the nasopharynx and the sinuses. There is also limited evidence that formaldehyde may cause cancer at other organ sites.[1,8] The current OSHA permissible exposure level is 0.75 ppm for an 8-hour time-weighted average and 2 ppm for a 15-minute short-term exposure.

Formaldehyde has been measured at the fire scene by Lowry et al.,[41] Brandt-Rauf and colleagues,[11] and Jankovic et al.[37] Lowry et al. reported combined formaldehyde and acetaldehyde levels, with a mean of 5 ppm and a range of 1 to 15 ppm.[41] Brandt-Rauf and colleagues found aldehydes, including formaldehyde, at 4 of 14 fires at concentrations of 0.1 to 8.3 ppm.[11] Jankovic et al. detected formaldehyde at levels up to 8 ppm during knockdown and only 0.4 ppm during overhaul.[37] They also reported that airborne concentrations of formaldehyde inside the mask ranged from nondetectable to 0.3 ppm.

Diesel Exhaust

Considerable experimental and epidemiologic evidence gathered over the past 15 years suggests that constituents of diesel exhaust emissions are carcinogenic and may present a risk to occupations with regular exposure. Firefighters have significant potential for exposure to diesel exhaust, because fire trucks with diesel engines are routinely started inside of and backed into firehouses.

Froines and colleagues studied the concentration of diesel exhaust particulates in the air inside firehouses in New York, Boston, and Los Angeles in 1985.[24] Participating firefighters wore personal air samplers throughout the work shift while they were in the firehouse.

Unlike studies of air contaminants at the fire scene, the concentrations of airborne diesel particulate measured in this study should accurately reflect the actual exposure of firefighters to diesel emissions. Firefighters obviously do not wear respiratory protection at the firehouse. In addition, firefighters spend much of the work shift inside the firehouse, so that the 8-hour time-weighted average concentration reported by Froines et al. should meaningfully approximate the diesel exhaust exposure of urban firefighters on the job.[24]

Significant exposure to diesel exhaust particulates was detected.[24] Total airborne particulates from diesel exhaust emissions ranged from 170 to 480 $\mu g/m^3$. Worst case scenario sampling, during which a very active shift was simulated, detected levels of diesel exhaust particulates in the air of fire houses as high as 748 $\mu g/m^3$. The authors conclude that these levels of diesel exhaust emissions may be associated with a significant carcinogenic risk and efforts to reduce exposure should be made.[24] Unlike exposures received at the fire scene, diesel exhaust emissions emanate from a specific source that can be controlled with local ventilation attached to the exhaust pipe of the fire truck.

Other Agents

Although less well studied, there are additional environmental agents to which firefighters are exposed and for which experimental and/or epidemiologic studies support a relationship between exposure to the agent and the development of cancer. Examples include polychlorinated biphenyls (PCBs), various furans, styrene, and methylene chloride. In the studies by Jankovic et al.[37] and Lowry et al.[41] discussed above, the latter three agents or groups of agents were found in measurable concentrations at multiple fires, but data on actual airborne levels were not provided. Indeed, in the study by Lowery and colleagues, 70 organic agents were repeatedly identified in the smoke at multiple fires in Dallas.[41] Given the large number of chemicals that have been identified as being carcinogenic in the past two decades, at least in rodent test systems,[55] it is likely that fire smoke contains additional carcinogens beyond those identified to date.

Conclusion

In conclusion, empirical data are now sufficient to support the notion that firefighters are exposed to carcinogens in their work environment. The significance of such exposures is still unresolved. The exposures of firefighters are intermittent and variable in intensity. The respiratory protection they use is of uncertain efficacy and limited acceptability in the real world. Important exposures such as asbestos and diesel exhaust may occur during overhaul or at the firehouse, when respirators are not typically used. Furthermore, even if the dose of various carcinogens received by firefighters were better known, the residual uncertainty about the degree of risk imparted would be great. Although the fact that firefighters are exposed to carcinogens in their work environment has been established, much additional work remains to be done. Sufficient knowledge exists at present, however, to justify diligent efforts to reduce the exposure of firefighters to known carcinogenic agents.

PREVALENT CANCERS IN FIREFIGHTERS AND ASSOCIATIONS WITH CARCINOGENIC OCCUPATIONAL EXPOSURES

The results of 19 epidemiologic studies of cancer in firefighters published in the medical literature are summarized below. The data show that employment as a firefighter increases the risk of developing and dying from certain specific cancers: leukemia, nonHodgkin's lymphoma, multiple myeloma, and cancers of the brain, urinary bladder, and, possibly, prostate, large intestine, and skin. Graphic presentations of data related to these specific cancers (Fig. 1–6) include results from all published epidemiologic studies of firefighters that reported on that cancer. (Results for nonspecific organ systems or sites, e.g., digestive system or hematopoietic/lymphatic system, were not included.) For

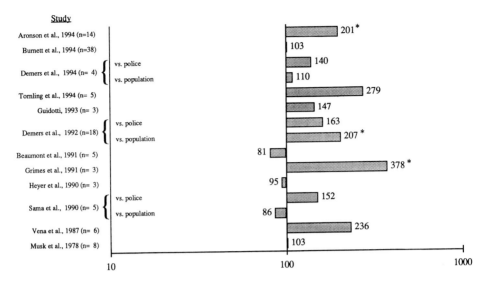

FIGURE 1. Brain cancer risk estimates for firefighters from published epidemiologic studies. Studies listed by first author and publication year (n = observed number of cancers among firefighters). Risk ratio expressed by authors as SMR, PMR, SIR, or RR, with null value (no excess risk) equaling 100 on log^{10} scale. *Statistically significant increase in risk ratio (p<0.05).

a given study, the "risk ratio" reported is the measure the authors used to express the association between firefighting and cancer: a standardized mortality ratio (SMR), proportionate mortality ratio (PMR), standardized incidence ratio. (SIR), or a relative risk, incidence density ratio or odds ratio multiplied by 100 (RR). The number of cancer cases or deaths observed among firefighters, the risk ratio, and the statistical significance of the result are indicated for each study. Unless otherwise stated, the reference group used to calculate a risk ratio was the general population; certain studies calculated risk ratios for more than one reference group, for example, police officers and the general population.

Brain Cancer

Chemical exposures that are suspected causes of brain tumors include vinyl chloride, benzene, PAHs, PCBs, N-nitroso compounds, triazenes and hydrazines.[36,65,71] Recent epidemiologic studies consistently have found that brain cancer is strongly associated with firefighting, as shown in Figure 1. Generally, excess risk was most notable within 15–30 years of exposure, i.e., after a relatively short latency.[2,16,68,70] Howe and Burch[34] analyzed all cancer mortality studies of firefighters available as of 1989 and concluded that brain cancer fulfilled the criteria indicative of a causal association with firefighting, with a pooled SMR of 143 (95% confidence interval =93–212).

A study by Aronson et al.[2] of firefighters in metropolitan Toronto reported a statistically significant overall SMR of 201 (95% CI=110–337) for brain cancer, with the highest mortality among those with 5–9 years duration of employment as a firefighter (SMR=625, 95% CI=170–1,600). Demers et al.[16] analyzed mortality data from three northwestern cities in the United States and found that firefighters with 10–19 years of employment were at greatest risk (SMR=353, 95% CI=150–700). Although based on only three deaths, an analysis of Honolulu firefighters by Grimes et al.[28] found a PMR of 378 (95% CI=122–1,171) for brain and other central nervous system cancers; analyses by years of employment, were not reported. Tornling et al.[68] were unique in finding dose-response relationships between brain cancer incidence and increasing age, duration of employment, and years since hire, and between brain cancer mortality and increasing age, duration of employment, and estimated number of fires fought among Stockholm firefighters who worked during 1931–1983.

Cancers of Hematopoietic and Lymphatic Systems

Leukemia and lymphoma are associated with environmental and occupational exposure to benzene and 1,3-butadiene.[36,47,49,72] The prevalence of benzene as a solvent, as a component of gasoline, and as a combustion product that forms during the burning of plastics and synthetics, and of 1,3-butadiene, a monomer found in tires and synthetic rubber products, guarantees that firefighters will be exposed to the gases released by these materials as they burn. Chemical exposures that have been associated with multiple myeloma include benzene and petroleum products. Multiple myeloma risk is also increased in farmers, paper producers, furniture manufacturers, and woodworkers.[9]

Leukemia

As seen in Figure 2, the majority of epidemiologic studies have found that firefighters are at increased risk of leukemia.[2,22,33,50,59] For example, Feuer and Rosenman[22] reported a statistically significant PMR of 276 for firefighters compared to police officers in New Jersey and an almost twofold increase in mortality compared to the general population in New Jersey and in the United States. Similarly, Sama et al.[59] found that firefighters had almost three times the risk of police officers when incident cases reported to the Massachusetts Cancer Registry from 1982 to 1986 were examined (age-standardized mortality odds ratio=267, 95% CI=62–1,154). Several studies found that the highest risk occurred at older ages, after at least 30 years latency or duration of employment.[2,16,33] However, a recent large study from NIOSH[12] combining mortality data from 27 states reported excess risk for firefighters younger than 65 (PMR=171, 95% CI=118–240).

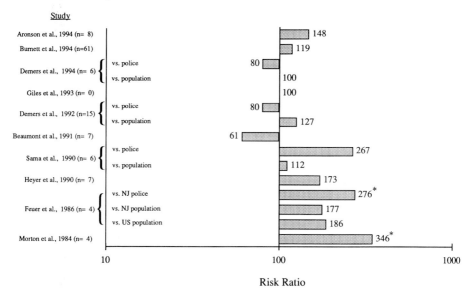

FIGURE 2. Leukemia risk estimates for firefighters from published epidemiologic studies. Studies listed by first author and publication year (n = observed number of cancers among firefighters). Risk ratio expressed by authors as SMR, PMR, SIR, or RR, with null value (no excess risk) equaling 100 on $\log^{10}$ scale. *Statistically significant increase in risk ratio (p<0.05).

NonHodgkin's Lymphoma

Several studies of firefighters evaluated this group of malignant diseases. Without exception, marked increases in risk were found (data not shown).[2,12,15,26,59] The study from the Massachusetts Cancer Registry by Sama et al. found a statistically significant SMOR of 327 (95% CI=119–898) for firefighters relative to police officers.[59] Studies by Giles et al.[26] from Melbourne, Australia, and Aronson et al.[2] from Toronto, Canada, reported that firefighters had twice the risk of non-Hodgkin's lymphoma of males in the general population.

Multiple Myeloma

Few individual epidemiologic studies of firefighters had sample sizes sufficient to assess risk of multiple myeloma (data not shown). Two of the four published studies that included multiple myeloma found lower than expected risk, based on one[2] or two[15] cases among firefighters. Two other studies reported increased risk associated with firefighting.[12,33] Although the confidence intervals were wide, the analysis of a cohort of Seattle firefighters by Heyer et al.[33] reported an overall SMR of 225 (95% CI=47–660) and, for men with 30 years or more of fire combat duty, a statistically significant SMR of 989 (95% CI=120–3,571). Using the mortality experience for 1984–1990 for firefighters from 27 states, Burnett et al. found a statistically significant age-adjusted PMR of 148 (95% CI=102–207).[12] Howe and Burch[34] combined the results of all cancer mortality studies of firefighters available as of 1989 (including four unpublished reports) and concluded that there was consistent evidence of a causal association between multiple myeloma and firefighting (pooled SMR=151, 95% CI=91–235).

Cancers of Genitourinary System

Bladder Cancer

Occupational chemical exposures known to cause bladder cancer include several aromatic amines, solvents, benzidine, PAHs, coal tars and pitches, soot and oils,[13,31,36] substances commonly encountered by firefighters, particularly at fires in commercial establishments. As seen in Figure 3, the majority of epidemiologic studies found that firefighting was associated with increased risk for bladder cancer. Guidotti[29] and Vena et al.[70] both reported a threefold increase in bladder cancer deaths compared to general population rates, with peak risks for firefighters age 60 and older, with latency of 40 or more years. Using incident cases from the Massachusetts Cancer Registry, Sama et al.[59] found a statistically significant increased risk for firefighters compared to police officers (SMOR=211, 95% CI=107–414) and to the general population (SMOR=159, 95% CI=102–250). Demers et al.[16] reported, based on two deaths, that the rate of bladder cancer was markedly lower than expected in a cohort of firefighters employed at least one year between 1944 and 1979 in Seattle and Tacoma, Washington, and Portland, Oregon (SMR=23, 95% CI=3–83 compared to the general population; age-standardized incidence density ratio =16, 95% CI=2–124 compared to police officers). However, in a recent retrospective cohort study among the firefighters from Seattle and Tacoma, the authors determined that cancer incidence was greater than expected relative to both the general population and the police, based on 18 incident bladder cancer cases among firefighters reported to a Surveillance, Epidemiology and End Results (SEER) tumor registry during 1974–1989.[15]

Kidney Cancer

Occupational exposures that have been implicated as risk factors for renal cell carcinoma include asbestos, PAHs, lead phosphate, dimethyl nitrosamine, coke oven emis-

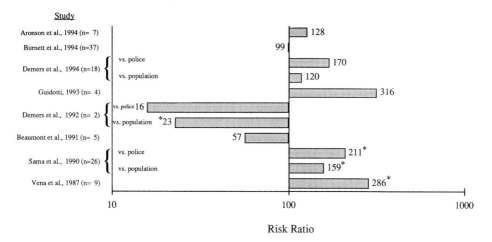

FIGURE 3. Bladder cancer risk estimates for firefighters from published epidemiologic studies. Studies listed by first author and publication year (n = observed number of cancers among firefighters). Risk ratio expressed by authors as SMR, PMR, SIR, or RR, with null value (no excess risk) equaling 100 on log^{10} scale. *Statistically significant increase in risk ratio (p<0.05).

sions, and gasoline.[36,56,62] This list clearly includes agents encountered in firefighting; however, the eight epidemiologic studies that assessed kidney cancer in firefighters did not show consistently elevated risk (data not shown). Burnett et al.[12] and Guidotti[29] did find statistically significant excess mortality among firefighters from 27 states in the United States and from Alberta, Canada, respectively. Guidotti's SMR of 414 (95% CI=166–853) for kidney and ureter cancer was the highest SMR reported in the study. Risk was greatest after 40–49 years latency and increased with duration of employment as a firefighter and with a calculated index of firefighting exposure opportunity.[29] Conversely, a number of studies have reported lower than expected risk among firefighters.[2,6,15,16] Studies from the northwestern United States by Demers and others found lower than expected kidney cancer mortality[16] and incidence.[15] Although based on only two deaths, the SMR of 27 (95% CI=3–97) for kidney cancer mortality was statistically significant relative to the general population.[16]

PROSTATE CANCER

High rates of prostate cancer have been reported among workers with cadmium exposure and in chemists, farmers, loggers, textile workers, painters, and rubber industry workers.[20,27,38,48] While no obvious carcinogenic exposure is common to all these groups, occupational risk factors clearly should be considered along with endocrinologic, sexual, and dietary factors in the etiology of prostate cancer. Figure 4 summarizes the data on firefighters' risk for prostate cancer. A 30–50% increase in risk was consistently found in the majority of studies. Giles et al.[26] found that prostate cancer incidence among firefighters employed in Melbourne, Australia, between 1917 and 1989 occurred at twice the expected rate (SIR=209, 95% CI=67–488). A proportionate mortality study by Grimes et al.[28] from Honolulu found statistically significant increases for prostate cancer in both Caucasian (PMR=370, 95% CI=171–802) and Hawaiian (PMR=335, 95% CI=107–1,045) firefighters. On the other hand, Beaumont et al.[6] found a statistically significant decrement in prostate cancer mortality (SMR=38, 95% CI=16–75) in a retrospective cohort study of firefighters employed between 1940 and 1979 in San Francisco.

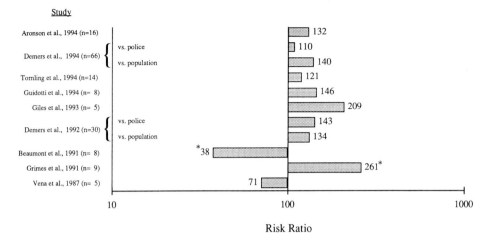

FIGURE 4. Prostate cancer risk estimates for firefighters from published epidemiologic studies. Studies listed by first author and publication year (n = observed number of cancers among firefighters). Risk ratio expressed by authors as SMR, PMR, SIR, or RR, with null value (no excess risk) equaling 100 on log^{10} scale. *Statistically significant increase in risk ratio (p<0.05).

TESTICULAR CANCER

Only two epidemiologic studies specifically addressed testicular cancer in firefighters.[2,26] Giles et al.[26] found no association between testicular cancer incidence and employment as a firefighter in Melbourne, Australia, between 1917 and 1989; however, this study was restricted to cancers that occurred between 1980 and 1989, and only two cases were reported. A recent report by Aronson et al.[2] found higher than expected mortality for men employed by the Toronto Fire Department during 1950–1989. Over this 40–year period, three testicular cancer deaths occurred in the cohort when only 1.19 were expected based on the Toronto male population of the same age and calendar period, for an overall SMR of 252 (95% CI=52–737). All three deaths occurred in younger men with less than 15 years as firefighters (SMR=366, 95% CI=75–1,069) and within 20 years of first exposure (SMR=326, 95% CI=67–953). The epidemiologic characteristics of testicular cancer show that it occurs most commonly from age 20 to 34, with a white:black ratio of 4:1 and a positive correlation with socioeconomic status.[60] The incidence and mortality rates in men younger than 30 have been increasing over time. Although occupational risk factors have not been studied well, exposures to solvents and paints have been implicated.[23] Testicular cancer risk should be assessed in future studies of firefighters.

Cancers of the Digestive System

Several established occupational exposures increase the risk of cancer of the digestive system: asbestos, cutting and lubricating oils, dyes, solvents, and metallic compounds.[25,36] It is hypothesized that, once cleared from the airways, inhaled particles and the carcinogens that adhere to them are transferred to the gastrointestinal tract and swallowed and exert their effect on the digestive epithelium. Cancers of the rectum, colon, liver, pancreas, stomach, and esophagus were assessed in the majority of epidemiologic studies, but too few studies included cancers of the buccal cavity or pharynx for meaningful discussion.

LARGE INTESTINE

Of particular relevance to firefighters are the higher than expected rates of colon and rectal cancer observed in workers with exposure to asbestos.[63] Figure 5 demonstrates that excess rectal cancer has been found consistently in many studies of firefighters.[2,6,12,15,52,59,68,70] A similar pattern was evident for colon, colorectal or "intestinal" cancer,[7,15,16,18,26,30,52,70] although the risk ratios tended to be somewhat lower (data not shown).

An analysis by Burnett and colleagues[12] of mortality data for firefighters from 27 states found a statistically significant excess of rectal cancer, particularly under age 65 (PMR=186, 95% CI=110–294). Orris et al.[52] reported significantly higher mortality in Chicago firefighters during 1940–1988 for both rectal (PMR=164, 95% CI=114–230) and colon (PMR=131, 95% CI=104–165) cancers. In three other studies,[2,68,70] rectal cancer mortality among firefighters occurred at twice the expected rate, but these results did not reach statistical significance. Slightly lower than expected mortality was observed in two analyses of firefighters from the northwestern United States.[16,33] However, the latest study from this area found that rectal cancer incidence was similar to both the police and the general population, while colon cancer incidence, although not significantly elevated, appeared to increase with duration of employment as a firefighter.[15]

LIVER CANCER

Primary liver cancer is rare in the general population of the United States. Angiosarcoma of the liver has been associated with occupational and environmental exposures, including arsenic and vinyl chloride monomer from PVC.[21,36] PVC can be assumed to be present at every structural fire site in recent years involving furniture, electrical wire, and cable insulation and water pipes, and at automobile fires.

Five epidemiologic studies reporting results for cancer of the liver (including

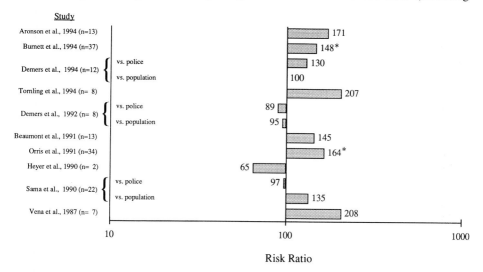

FIGURE 5. Rectal cancer risk estimates for firefighters from published epidemiologic studies. Studies listed by first author and publication year (n = observed number of cancers among firefighters). Risk ratio expressed by authors as SMR, PMR, SIR, or RR, with null value (no excess risk) equaling 100 on log[10] scale. *Statistically significant increase in risk ratio (p<0.05).

cancer of the biliary passages and gallbladder) were all based on small numbers of cases observed in firefighters (data not shown). The study with the largest number[6] found a twofold excess for liver cancer mortality relative to the United States population among firefighters in San Francisco who were employed between 1940 and 1970 (SMR=191, 95% CI=87–363, n=9). Tornling et al.[68] found a nonsignificant increase in mortality (SMR=149, 95% CI=41–381, n=4) but a slight decrement in incidence (SMR=85, 95% CI=23–218, n=4) for liver cancer in Stockholm firefighters employed during 1931–1983, relative to regional rates. Three additional studies found no association between firefighting and liver cancer.[2,16,70] Although such an association is biologically plausible, only a very large study or meta-analysis would have adequate statistical power to detect an increase in this rare cancer.

PANCREATIC CANCER

Many occupations and chemical carcinogens have been studied in relation to pancreatic cancer, with little consensus.[53] Workers in chemical, petroleum, and metallurgic industries may have particularly high risk from exposures such as benzidine, β-naphthylamine derivatives, and metal dusts.[40,53,54] In general, epidemiologic data suggest that firefighting is not associated with cancer of the pancreas (data not shown). One study found a large but nonsignificant increase in incidence for firefighters compared to police officers (SMOR=319) but not compared to the general population (SMOR=98) in Massachusetts.[59] Eight additional investigations assessed pancreatic cancer in firefighters: one study reported a nonsignificantly decreased risk (SMR=38),[26] three studies reported slightly elevated risk,[2,6,30] and four studies reported equal risk relative to the general population.[15,16,26,68]

STOMACH AND ESOPHAGEAL CANCER

Adenocarcinoma of the stomach and cancer of the esophagus have been associated with asbestos exposure;[10,25,62] as discussed above, asbestos is prevalent at the majority of structural fires. Workers involved in rubber manufacturing, metal working, wood and paper working, and coal mining have also shown high rates of stomach cancer.[25]

Most of the epidemiologic studies that addressed stomach cancer found a positive association with firefighting,[6,15,16,18,33,68,70] but none of the overall results were statistically significant (data not shown). Eliopulos et al.[18] studied a cohort of firefighters employed during 1939–1978 by the Western Australia Fire Brigade. Mortality from stomach cancer was increased twofold relative to the general population (PMR=202, 95% CI=65–470). A study of firefighters employed in Stockholm during 1931–1983 found a small overall SMR of 121 for stomach cancer mortality;[68] however, both incidence and mortality increased with duration of employment and number of fires fought. Although tests for trend did not reach statistical significance, stomach cancer incidence was significantly elevated for firefighters with more than 30 years employment (SMR=289, 95% CI=149–505) or who fought more than 1,000 fires (SMR=264, 95% CI=136–461).

The data for cancer of the esophagus are more equivocal. Equal numbers of studies found positive[6,15,70] and negative[2,16,33] associations with firefighting (data not shown). Beaumont et al.[6] found that mortality from esophageal cancer occurred at twice the expected rate (SMR=204, 95% CI=105–357) in a retrospective cohort study of firefighters employed between 1940 and 1979 in San Francisco. No increase was demonstrated with increasing duration of employment or latency—in fact, the highest rate was seen for those with less than 20 years as a firefighter. The authors postulate that an interaction between smoke exposure and alcohol consumption could explain the pattern of cancer mortality in their study population: elevated rates for cancers of the liver, esophagus, buccal cavity, and pharynx.

Skin Cancer

Skin cancer is a heterogeneous group of diseases, the majority of which are malignant melanoma (30,000 new cases in the United States per year) or basal cell or squamous cell carcinomas (500,000 new cases per year). The most common risk factor for cancers of the skin is prolonged and intense exposure to sunlight. Occupational exposure to soot and tars, coke oven emissions, arsenic, and cutting oils also have been associated with increased risk.[19,36] Substances containing carcinogenic agents such as PAHs and PCBs may be absorbed by the skin of exposed body areas, including the hands, arms, face and neck, and other sites when protective clothing is permeated. Contact with these substances can occur during fire knockdown and overhaul and during the cleaning of clothing or equipment.

Figure 6 summarizes the studies that addressed skin cancer risk. (In studies that failed to differentiate melanoma from non-melanoma skin cancer, mortality rates are likely to include only melanoma since other forms of skin cancer are rarely fatal.) Several studies found that firefighters had a statistically significant excess risk of skin cancer compared to the general population.[12,22,59] Using deaths reported to a retirement system between 1974 and 1980, Feuer and Rosenman[22] found an almost threefold increase in skin cancer mortality for New Jersey firefighters compared to the United States population (PMR=270, p<0.05); firefighters were at somewhat higher risk than the general New Jersey population (PMR=190) but at the same risk as New Jersey police officers (PMR=135). Risk among firefighters clearly increased with duration of employment and interval since first employment (PMR=388 for more than 25 years duration; PMR=314 for more than 27 years latency); it was not clear which referent population was used for these comparisons. Sama et al.[59] analyzed incident melanoma cases reported during 1982–1986 to the Massachusetts Cancer Registry. They found a statistically significant excess for firefighters in comparison to the state population (SMOR=292, 95% CI=170–503) but no excess in comparison to police officers except in the age group 55–74 years (SMOR=513, 95% CI=150–1,750). Howe and Burch[34]

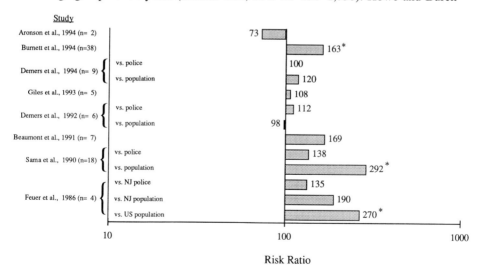

FIGURE 6. Skin cancer risk estimates for firefighters from published epidemiologic studies. Studies listed by first author and publication year (n = observed number of cancers among firefighters). Risk ratio expressed by authors as SMR, PMR, SIR, or RR, with null value (no excess risk) equaling 100 on $\log^{10}$ scale. *Statistically significant increase in risk ratio (p<0.05).

combined the results of the studies of cancer in firefighters published through 1989 and determined that there was evidence of a statistically significant increase in risk of melanoma (pooled SMR=173, 95% CI=103–274). However, they concluded that several criteria used to define a causal association were not fulfilled—for example, the ability to rule out potential confounders such as sunlight exposure and the limited evidence of a dose-response relationship.

Lung Cancer

As discussed above, firefighters may be routinely exposed to many known or suspected lung carcinogens, including asbestos, arsenic, PAHs, vinyl chloride and formaldehyde.[58] Inhalation exposure can occur during active fire combat as well as during the overhaul phase when protective breathing equipment is usually removed.

Accordingly, lung cancer was specified a priori in the majority of epidemiologic studies as an outcome that would be plausibly related to firefighting. Of the 16 published studies that addressed cancer of the respiratory tract, not one found a statistically significant excess risk of lung cancer for firefighters (data not shown). Only two cohort studies[29,32] found moderately increased risks: Guidotti from Canada, with an SMR of 142 (95% CI=91–211) for deaths occurring during 1927–1987, and Hansen et al. from Denmark, with an SMR of 163 for deaths occurring during 1970–1980 (95% CI=75–310). A case-control study using Missouri Cancer Registry cases diagnosed between 1980 and 1985 found the category that included police, firefighters, and protective service occupations had elevated risks for squamous-cell carcinoma, small-cell carcinoma, and other or mixed cell types, but not for adenocarcinoma of the lung.[73] These elevated risks were limited to current smokers only.

Discussion

These epidemiologic studies clearly demonstrate increased risk of several cancers that can be plausibly linked with carcinogenic exposures encountered by firefighters in their work. The data most strongly suggest that firefighters are at increased risk of developing and dying from leukemia, nonHodgkin's lymphoma, multiple myeloma, and cancers of the brain and bladder. The majority of studies that examined these cancers found markedly elevated risks for firefighters, and there are no viable alternative hypotheses or strong confounders that could readily explain their increased prevalence. Furthermore, exposure assessment studies have detected substances in the firefighting environment that are known or suspected causes of these cancers. Weaker but still plausible evidence links firefighting to increased risk of rectal, colon, stomach, and prostate cancers and melanoma.

The limitations of the epidemiologic data must be acknowledged. Most of the studies examined relatively small populations of firefighters and thus have low statistical power to analyze rare tumors. To increase their sample size, many of the studies analyzed deaths occurring over several decades; this technique introduces problems related to (a) trends in diagnoses, (b) differences in exposure over time, since many potential carcinogens, such as chemicals and synthetic materials, were introduced at different times during the relevant exposure periods, and (c) changes in protective equipment and awareness of hazards. Limited documentation of exposure is also a problem. Some studies relied on occupation as recorded on a death certificate or tumor registry, which may reflect the current or most recent job instead of the usual occupation. Recent studies have examined risk in relation to duration of active fire combat duty, latency (years since hire), age at diagnosis (active duty versus retirement), and number of fires fought. However, none were able to rank firefighters according to a cumulative index incorporating intensity of exposure. As a result, heavily exposed firefighters are comingled with lightly exposed firefighters, and the risks to the heavily exposed firefighters are diluted out and underestimated by the design of the studies.

None of the epidemiologic studies were able to take into account potential confounding variables other than age that could explain the observed associations between firefighting and cancer.[42] It is unlikely, however, that increased mortality rates among firefighters can be attributed solely to the personal lifestyle factors—diet, alcohol intake, cigarette smoking—that have been linked with certain cancers. The vast majority of studies found no excess risk of lung cancer, suggesting that firefighters are not more likely to smoke than the general population or other protective service workers. In fact, surveys have found that the proportion of firefighters who smoke is similar to the proportion of other service and blue collar workers who smoke.[5,59,67] In studies of occupation and cancer that did collect information on lifestyle factors, most associations remained unchanged after controlling for cigarette smoking,[4,17] and biased attribution of cause of death among smokers compared to nonsmokers has been shown to overestimate associations between smoking and cancer.[66]

The latency period for most of the relevant cancers associated with exposure to chemical carcinogens is likely to be at least three or four decades. Therefore, studies to date have not had sufficient follow-up time to detect the full extent of occupational cancer in the firefighters at greatest risk—those who were increasingly exposed to chemical carcinogens throughout the 1940s, 1950s, and 1960s without the benefit of modern protective equipment or awareness of hazards.

The results of the studies also may be subject to the paradox of the healthy worker and survivor effects.[3,35,46] Healthy individuals are more likely than unhealthy persons to seek and gain employment and to remain in their jobs. This effect is amplified by the stringent initial screening process and good employment benefits associated with employment as a firefighter, as evidenced by their low all-cause mortality rates. Although the healthy worker effect has less impact on cancer than on other causes of death, the higher than expected rates of cancer mortality among firefighters in comparison to the general population and, in particular, to other workers are unsettling. Indeed, the shortcomings of the epidemiologic studies are more likely to dilute or mask associations between occupational exposures of firefighting and cancer than to create falsely positive associations.

Few of the results presented reached statistical significance, and the confidence intervals around the risk ratios were generally wide. Statistical significance is determined by the magnitude of the exposure-disease association, the accuracy or variability of the exposure and outcome measurements, and the size of the study population. Therefore, the small numbers of cancers observed in individual studies contribute to instability in the risk estimates. Future studies that are able to include not just deaths but all incident cancers from large cohorts will benefit from analyzing greater numbers of events. Figures 1–6 illustrate the preponderance of evidence implicating certain specific cancers associated with firefighting. Although these cancers warrant particular attention, future investigations should continue to cast a wide net that includes all relevant cancers. The downside of testing many outcomes in relation to a number of exposure variables is that some associations may appear to be statistically significant by chance alone.

Because most of the epidemiologic studies used the retrospective cohort study design, investigators had access to employer records regarding employment period, work assignments, and vital status, rather than just occupation as recorded on a death certificate. Attempts should be made in future studies, particularly those with prospective components, to develop measures of acute and cumulative exposures on an individual basis, although potential misclassification will always be a concern given the nature of the firefighting environment. The techniques of molecular biology increasingly are being used to develop biomarkers of exposure in occupational and environmental settings.

For example, Liou et al.[39] monitored two biomarkers in firefighters: sister chromatid exchange (SCE), a general indicator of genetic damage resulting from exposure to mutagens and carcinogens, and polycyclic aromatic hydrocarbon (PAH)-DNA adducts, which are thought to measure the initiation of carcinogenic changes associated with exposure to PAHs. After controlling for charcoal-broiled food consumption, cigarette smoking and race, firefighters had a statistically significant fourfold higher risk of detectable PAH-DNA adduct levels compared to unexposed controls. This association may be specific to urban, structural firefighting; a similar study in wildland firefighters in California found no association between forest fire activity and PAH-DNA adducts.[57] The incorporation of biologic markers of exposure, cancer susceptibility, and preclinical effects should be considered in future epidemiologic studies of firefighters.

Despite the limitations cited above, the available exposure assessment and epidemiologic studies present convincing and consistent evidence that the toxic exposures encountered in firefighting may increase the risk for certain specific cancers. The relatively high incidence rates with which some of these cancers occur (prostate, colon, rectum) and, for rarer cancers, the particularly strong association with firefighting or dismal survival probability (brain, multiple myeloma) underscore the importance of understanding and reducing the cancer risks attributable to firefighting.

REFERENCES

1. Acheson ED, Barnes HR, Gardner MJ, et al: Formaldehyde in the British chemical industry. Lancet 1:611–616, 1984.
2. Aronson KJ, Tomlinson GA, Smith L: Mortality among fire fighters in Metropolitan Toronto. Am J Ind Med 26:89–101, 1994.
3. Arrighi HM, Hertz-Picciotto I: The evolving concept of the health worker survivor effect. Epidemiology 5:189–196, 1994.
4. Axelson O: Aspects of confounding in occupational health. Scand J Work Environ Health 12:486–493, 1978.
5. Bates JT: Coronary artery disease in the Toronto fire department. J Occup Med 29:132–135, 1987.
6. Beaumont JJ, Chu GST, Jones JR, et al: An epidemiologic study of cancer and other causes of mortality in San Francisco firefighters. Am J Ind Med 19:357–372, 1991.
7. Berg JW, Howell MA: Occupation and bowel cancer. J Toxicol Environ Health 1:75–89, 1975.
8. Blair A, Stewart PA, Hoover RN: Mortality from lung cancer among workers employed in formaldehyde industries. Am J Ind Med 17:683–699, 1990.
9. Blattner WA: Multiple myeloma and macroglobulinemia. In Schottenfeld D, Fraumeni JF (eds): Cancer Epidemiology and Prevention. Philadelphia, WB Saunders, 1982, pp 795–813.
10. Botha JL, Irwig LM, Strebel PM: Excess mortality form stomach cancer, lung cancer, and asbestosis and/or mesothelioma in crocidolite mining districts in South Africa. Am J Epidemiol 123:30–40, 1986.
11. Brandt-Rauf PW, Fallon LF Jr, Tarantini T, et al: Health hazards of fire fighters: Exposure assessment. Br J Ind Med 45:606–612, 1988.
12. Burnett CA, Halperin WE, Lalich NR, Sestito JP: Mortality among fire fighters: A 27 state survey. Am J Ind Med 26:831–833, 1994.
13. Cole P, Hoover R, Friedell GH: Occupation and cancer of the lower urinary tract. Cancer 29:1250–1260, 1972.
14. Committee on Fire Toxicology: Fire and Smoke: Understanding the Hazards. Washington, DC, National Academy Press, 1986.
15. Demers PA, Checkoway H, Vaughan TL, et al: Cancer incidence among firefighters in Seattle and Tacoma, Washington (United States). Cancer Causes Control 5:129–135, 1994.
16. Demers PA, Heyer NJ, Rosenstock L: Mortality among firefighters from three northwestern United States cities. Br J Ind Med 49:664–670, 1992.
17. Dubrow R, Wegman DH: Setting priorities for occupational cancer research and control: Synthesis of the results of occupational disease surveillance studies. J Natl Cancer Inst 71:1123–1142, 1983.
18. Eliopulos E, Armstrong BK, Spickett JT, Heyworth, F: Mortality of fire fighters in Western Australia. Br J Ind Med 41:183–187, 1984.
19. Emmett EA: Occupational skin cancers. State Art Rev Occup Med 2:165–177, 1987.
20. Ernster VL, Selvin S, Brown SM, et al: Occupation and prostatic cancer. A review and retrospective analysis based on death certificates in two California counties. J Occup Med 21:175–183, 1979.

21. Falk H, Caldwell GG, Ishak KG, et al: Arsenic-related hepatic angiosarcoma. Am J Ind Med 2:43–50, 1981.
22. Feuer E, Rosenman K: Mortality in police and firefighters in New Jersey. Am J Ind Med 9:517–527, 1986.
23. Fleming L: Cancers of the reproductive organs. In Rosenstock L, Cullen MR: Textbook of Clinical Occupational and Environmental Medicine. Philadelphia, WB Saunders, 1994, pp 591–599.
24. Froines JR, Hinds WC, Duffy RM, et al: Exposure of fire fighters to diesel emissions in fire stations. Am Ind Hyg Assoc J 48:202–207, 1987.
25. Frumpkin H: Cancer of the liver and gastrointestinal tract. In Rosenstock L, Cullen MR: Textbook of Clinical Occupational and Environmental Medicine. Philadelphia, WB Saunders, 1994, pp 576–584.
26. Giles G, Staples M, Berry J: Cancer incidence in Melbourne metropolitan fire brigade members, 1980–1989. Health Rep 5:33–38, 1993.
27. Greenwald P: Prostate. In Schottenfeld D, Fraumeni JF (eds): Cancer Epidemiology and Prevention. Philadelphia, WB Saunders, 1982, pp 938–946.
28. Grimes G, Hirsch D, Borgeson D: Risk of death among Honolulu fire fighters. Hawaii Med J 50:82–85, 1991.
29. Guidotti TL: Mortality of urban firefighters in Alberta: 1927–1987. Am J Ind Med 23:921–940, 1993.
30. Guidotti TL, Clough VM: Occupational health concerns of firefighting. Annu Rev Public Health 13:151–171, 1992.
31. Gustavsson P, Gustavsson A, Hogstedt C: Excess of cancer in Swedish chimney sweeps. Br J Ind Med 45:777–781, 1988.
32. Hansen ES: A cohort study on the mortality of firefighters. Br J Ind Med 47:805–809, 1990.
33. Heyer N, Weiss NS, Demers P, Rosenstock L: Cohort mortality study of Seattle fire fighters: 1945–1983. Am J Ind Med 17:493–504, 1990.
34. Howe GR, Burch JD: Fire fighters and risk of cancer: An assessment and overview of the epidemiologic evidence. Am J Epidemiol 132:1039–1050, 1990.
35. Howe GR, Chiarelli AM, Lindsay J: Components and modifiers of the healthy worker effect: Evidence from three occupational cohorts and implications for industrial compensation. Am J Epidemiol 128:1364–1375, 1988.
36. International Agency for Research on Cancer: IARC Monographs on the Evaluation of Carcinogenic Risks to Humans, Suppl 7, Overall Evaluations of Carcinogenicity: An Updating of IARC Monographs. IARC Lyon, France, 1987.
37. Jankovic J, Jones W, Burkhart J, Noonan G: Environmental study of fire fighters. Ann Occup Hyg 35:581–602, 1991.
38. Kipling MD, Waterhouse JAH: Cadmium and prostate cancer. Lancet 1:730–731, 1967.
39. Liou SH, Jacobsen-Kram D, Poirier MC, et al: Biological monitoring of fire fighters: Sister chromatid exchange and polycyclic aromatic hydrocarbon-DNA adducts in peripheral blood cells. Cancer Res 49:4929–4935, 1989.
40. Lin RS, Kessler II: A multifactorial model for pancreatic cancer in man. JAMA 245:147–152, 1981.
41. Lowry WT, Juarez L, Petty VCS, Roberts B: Studies of toxic gas production during actual structural fires in the Dallas area. J Forensic Sci 30:59–71, 1985.
42. Mahaney FX: Studies conflict on fire fighters' risk of cancer. J Natl Cancer Inst 83:908–909, 1991.
43. Markowitz S, Garibaldi K, Lilis R, Landrigan PJ: Asbestos exposure and fire fighting. Ann N Y Acad Sci 643:573–576, 1992.
44. Mastromatteo E: Mortality in city firemen. II. A study of mortality in firemen of a city fire department. Arch Ind Health 20:227–233, 1959.
45. McDiarmid MA, Lees PSJ, Agnew J, et al: Reproductive hazards of fire fighting. II. Chemical hazards. Am J Ind Med 19:447–472, 1991.
46. McMichael AJ: Standardized mortality ratios and the "healthy worker effect:" Scratching beneath the surface. J Occup Med 18:165–168, 1976.
47. McMichael AJ, Spirtas R, Kupper LL, Gamble JF: Solvent exposure and leukemia among rubber workers: An epidemiologic study. J Occup Med 17:234–239, 1975.
48. Monson RR, Fine LJ: Cancer mortality and morbidity among rubber workers. J Natl Cancer Inst 61:1047–1053, 1978.
49. Monson RR, Nakano KK: Mortality among rubber workers. Am J Epidemiol 103:284–296, 1976.
50. Morton W, Marjanovic D: Leukemia incidence by occupation in the Portland-Vancouver metropolitan area. Am J Ind Med 6:185–205, 1984.
51. Musk AW, Monson RR, Peters MJ, Peters RK: Mortality among Boston firefighters, 1915–1975. Br J Ind Med 35:104–108, 1978.
52. Orris P, Kahn G, Melius J: Mortality study of Chicago firefighters [abstract]. Revue D'Epidemiologie Et De Sante Publique 40 (Suppl 1):S90–91, 1992.
53. Partanen T, Kauppinen T, Degerth R, et al: Pancreatic cancer in industrial branches and occupations in Finland. Am J Ind Med 25:851–866, 1994.

54. Pietri F, Clavel F: Occupational exposure and cancer of the pancreas. A review. Br J Ind Med 48:583–587, 1991.
55. Rall DP, Hoga MD, Huff JE, et al: Alternatives to using human experience in assessing health risks. Annu Rev Public Health 8:355–3385, 1987.
56. Redmond CK, Ciocco A, Lloyd JW, et al: Longterm mortality study of steel workers. VI. Mortality from malignant neoplasms among coke oven workers. J Occup Med 14:621–629, 1972.
57. Rothman N, Correa-Villasenor A, Ford DP, et al: Contribution of occupation and diet to white blood cell polycyclic aromatic hydrocarbon-DNA adducts in wildland firefighters. Cancer Epidemiol Biomarkers Prev 2:341–347, 1983.
58. Russi MB, Cone JE: Malignancies of the respiratory tract and pleura. In Rosenstock L, Cullen MR: Textbook of Clinical Occupational and Environmental Medicine. Philadelphia, WB Saunders, 1994, pp 543–555.
59. Sama SR, Martin TR, Davis L, Kriebel D: Cancer incidence among Massachusetts firefighters: 1982–1986. Am J Ind Med 18:47–54, 1990.
60. Schottenfeld D, Warshauer ME: Testis. In Schottenfeld D, Fraumeni JF (eds): Cancer Epidemiology and Prevention. Philadelphia, WB Saunders, 1982, pp 947–957.
61. Selikoff IJ, Hammond EC: Asbestos-associated disease in United States shipyards. CA Cancer J Clin 28:87–99, 1978.
62. Selifkoff IJ, Hammond EC, Seidman H: Mortality experience of insulation workers in the United States and Canada, 1943–1976. Ann N Y Acad Sci 330:91–116, 1979.
63. Selikoff IJ, Lee D: Asbestos and Disease. London, Academic Press, 1978.
64. Siemiatycki J: Risk Factors for Cancer in the Workplace. Boca Raton, FL, CRC Press, 1991.
65. Sinks T, Steele G, Smith AB, et al: Mortality among workers exposed to polychlorinated biphenyls. Am J Epidemiol 136:389–398, 1992.
66. Sterling TD, Rosenbaum WL, Weinkan JJ: Bias in the attribution of lung cancer as a cause of death and its possible consequences for calculating smoking-related risks. Epidemiology 3:11–16, 1992.
67. Sterling TD, Weinkan JJ: Smoking characteristics by type of employment. J Occup Med 18:743–754, 1976.
68. Tornling G, Gustavsson P, Hogstedt C: Mortality and cancer incidence in Stockholm fire fighters. Am J Ind Med 25:219–228, 1994.
69. Treitman RD, Burgess WA, Gold A: Air contaminants encountered by fire fighters. Am Ind Hyg Assoc J 41:796–802, 1980.
70. Vena JE, Fiedler RC: Mortality of a municipal-worker cohort: IV. Fire fighters. Am J Ind Med 11:671–684, 1987.
71. Waxweiler RJ, Stringer W, Wagoner JK, et al: Neoplastic risk among workers exposed to vinyl chloride. Ann N Y Acad Sci 271:40–48, 1976.
72. Young N: Benzene and lymphoma. Am J Ind Med 15:495–498, 1989.
73. Zahm SH, Brownson RC, Chang JC, Davis JR: Study of lung cancer histologic types, occupation and smoking in Missouri. Am J Ind Med 15:565–578, 1989.

JAMES M. MELIUS, MD, DrPH

CARDIOVASCULAR DISEASE AMONG FIREFIGHTERS

Center to Protect Workers' Rights
Washington, DC

Reprint requests to:
James M. Melius, MD, DrPH
Scientific and Medical Director
Center to Protect Workers' Rights
111 Massachusetts Ave.
Washington, DC 20001

Heart disease has long been recognized as a significant health risk for firefighters. State and local retirement systems for firefighters in many parts of North America have presumed that heart disease in an active firefighter has resulted from firefighting duties. Currently, 36 states have some type of heart presumption disability law for firefighters.[16] Most of these laws are more than 30 years old and predate much of our current scientific literature on firefighting. However, older publications on the risks of firefighting by physicians actively involved in providing care for this profession indicate strong support for the need for these laws.[33]

Over the past 20 years, more studies have been conducted on health risks for firefighters. This chapter reviews these studies as they relate to the risk of cardiovascular disease among firefighters.

FIREFIGHTING

To better elucidate the relationship between work as a firefighter and the risk of cardiovascular disease, several aspects of the firefighting environment are reviewed, including firefighters' exposures, the physical requirements of the work, and psychological factors.

Exposures

Firefighters may be exposed to a multitude of toxic substances when fighting a fire or performing other firefighting duties (see chapter 1). Carbon monoxide is the most commonly documented exposure and the exposure with the most direct cardiac toxicity, including increased cardiovascular disease mortality among occupational cohorts with prolonged exposure.[40] For example, a large study of Boston firefighters found that in approximately 15% of the individual fire exposures studied, carbon monoxide exposures exceeded 400 ppm and reached as high as 5,000 ppm.[44] Several other studies have documented elevated levels of carboxyhemoglobin in firefighters.[17,31,35,41,42]

Cyanide is another potential fire smoke contaminant with cardiac toxicity. Although cyanide exposure appears to be low in most fires,[5,17,25,44] occasional elevated levels may be found, especially if certain plastics are burning. A number of respiratory irritants are also commonly found in fire smoke, including particulates, acrolein, hydrogen chloride, nitrogen dioxide, polynuclear aromatic compounds, and benzene. Of these, acrolein is the most commonly measured at levels sufficient to produce respiratory irritation. In the Boston study, approximately 10% of firefighters were exposed at levels above 3 ppm.[44] Although these substances do not have any direct cardiac or metabolic toxicity, exposures may lead to significant short-term decreases in pulmonary function.[37] This in turn may increase the physiologic demands on the cardiovascular system.

The increasingly routine use of self-contained breathing apparatus (SCBA) by firefighters has decreased the risk of high exposures to these substances. However, in many fire suppression activities, it is not possible for firefighters to wear SCBA the entire time, either due to the physiologic stress from prolonged wear or from the lack of readily available refills for the SCBA.[17] Thus, significant exposures to fire smoke continues for most firefighters. Incomplete protection from the SCBA either due to pressure differentials or other problems also may contribute to this incomplete protection.

Physical Requirements

In evaluating the risks of cardiovascular disease, the physical demands of firefighting must be assessed. These physical requirements must be understood in the context of usual firefighting duties. A significant proportion of their work day is spent at rest or in light work in the fire station. However, at any time, including while asleep, firefighters may be called for rapid transport to the scene of a fire and quick engagement in the strenuous physical activities of firefighting.

Actual firefighting activities are strenuous and involve high muscular strength and aerobic capacity that may persist for several hours.[6,19,22,25] These activities often require firefighters to work at near maximal heart rates for long times.[1] The increase in heart rate appears to be initiated with responding to an alarm and thus may occur more often than just with actual fire suppression duties.[1,15] The high heart rate persists through the course of the fire suppression activities,[25] leaving little reserve for unexpected situations such as personal or victim rescue.

Several studies have attempted to evaluate the energy costs of different firefighting activities, usually by simulating these conditions.[12] Climbing stairs, climbing an aerial ladder, dragging hoses, and the simulated victim search and rescue had high energy costs. For example, one study of firefighters climbing stairs in full equipment found that they reach 80% of their maximum oxygen consumption and 95% of their maximal heart rate and required at least 39 ml of oxygen per kg per minute.[25] Other studies have shown similar requirements.[22]

Heat is another important determinant of the physical requirements for firefighting. High temperatures are an obvious part of the fire environment and place an additional burden on the firefighter. Studies indicate much greater physiologic demands on firefighters exposed to high heat although these are difficult to quantitate separate from the physical activities associated with firefighting tasks.[34] Studies of stimulated firefighting activities document the increased physiologic stress from the combination of exertion and high environmental temperatures, including a significant rise in core body temperature even with rest breaks.[8,34,38]

In addition to the stress from firefighting tasks and high temperatures, the use protective equipment creates an additional burden. Firefighters typically wear about 35 pounds of protective equipment, including boots, helmet, turn-out coats, and SCBA.[16] In addition to the additional weight, the typical equipment may interfere with the thoracic mechanics of breathing and leads to a greater retention of body heat under exercise and simulated firefighting activities.[22,32,34,45]

Psychological Stress

There are relatively few studies of stress among firefighters and none that directly relate to the issue of long-term risk for cardiovascular disease. A few studies have found physiologic evidence of stress in active-duty firefighters as measured through methods such as catecholamine excretion.[9,18] Several studies have focused on posttraumatic stress reactions among firefighters.[12] However, for an occupation that would appear to experience high levels of stress during fire and rescue operations, relatively few studies document this stress.

HEALTH STUDIES

The direct medical data available to assess the association between firefighting and cardiovascular disease includes a number of mortality studies, a few cross-sectional studies, and one small prospective study.

Mortality Studies

There are several cohort or proportionate mortality studies of firefighters that compare the cardiovascular disease mortality experience of firefighters with the general population. One must be careful in reviewing these studies to recognize the limitations of this method for evaluating the risk for cardiovascular disease in an occupational cohort.

The major limitation is the so-called "healthy worker effect." People in the workforce are usually selected in such a way that they have better health (and hence a lower mortality rate) than the general population usually used for comparison purpose in these studies.[28,39] This effect is usually expressed with the finding of a significantly lower standardized mortality ratio for cardiovascular disease and other diseases than the general public (which constitutes the comparison group for the study). For firefighters, this healthy worker effect should be particularly strong because of the selection of entry-level firefighters on the basis of their physical fitness. Unfortunately, there is no direct and universally accepted method for controlling for this effect when conducting a study or when comparing different studies. Comparing mortality from heart disease in firefighters to that in another employed cohort can be helpful but is often limited by the small size of the comparison population and the possibility of increased cardiovascular disease risk among the other comparison group.

Another important consideration is that most mortality studies do not completely ascertain the incidence of cardiovascular disease among the working population. Only cardiovascular disease deaths are recorded. Significant heart disease (i.e., previous myocardial infarction) is often not recorded on the death certificate. While this also applies for the comparison population, it can still lead to a significant underestimation of the magnitude of cardiovascular disease in the study group. Finally, there may be problems with coding for cardiovascular disease deaths because several different categories are involved.

Despite these limitations, some mortality studies of firefighter cohorts have documented an increased risk of mortality among firefighters. An early study of Toronto city firefighters (1921–1953) found a significantly increased risk of cardiovascular-renal disease mortality (SMR=1.41).[26] This increase was especially pronounced among firefighters ages 50–59. A later study assessing Toronto city firefighters hired from 1949 to

1959 and followed through 1984 found a statistically significant increased incidence of coronary artery disease mortality among firefighters (SMR = 1.73); this excess was most pronounced among firefighters 45 to 54 years old.[4]

A more recent study of metropolitan Toronto area firefighters found a slightly elevated (but not statistically significant) mortality from cardiovascular disease in this cohort, who worked in these fire departments from 1950 to 1989.[20] Interestingly, this study found a twofold increase in mortality from aortic aneurysm in this cohort (SMR = 2.26).

A small study in Hawaii and one in Connecticut found significantly increased mortality ratios for cardiovascular disease among firefighters, and proportionate mortality ratio studies in New Jersey and Chicago had similar findings.[10,11,30,36] Three other recent studies have found nonsignificant increases in cardiovascular disease mortality among firefighters.[13,14,27] On the other hand, other studies have found significantly decreased cardiovascular disease mortality among firefighters in Boston and in Sweden.[29,43]

Other Medical Studies

Barnard and colleagues in the Los Angeles area conducted some cross-sectional studies of firefighters to evaluate the occurrence of cardiovascular disease.[1–3] These studies reported findings of "ischemic heart disease" with normal coronary arteries among a small group of firefighters selected because of ischemic findings on near-maximal stress testing.[2] In addition, firefighters had a slightly higher rate of ischemic findings on near-maximal stress testing than a comparison group of underwriters, despite having a lower rate of other coronary heart disease risk factors than the comparison group.[3] The small number of firefighters studied and the obvious limitations of the study design restrict the applicability of these findings.

Although a prospective study of firefighters evaluating other cardiovascular disease risk factors in addition to the role of firefighting could add significant scientific information, such a study is difficult to conduct and expensive. As part of another prospective study in Boston, the cardiovascular disease experience of a small group of firefighters was evaluated.[7] This study of 171 Boston firefighters found no increase in the rate of cardiovascular disease events in this group over a 10-year period. However, members of the group participating in the study were screened for the presence of heart disease at the time of enrollment, thus eliminating firefighters who may have left work because of heart disease.

DISCUSSION

The limitations of the current methodology for conducting occupational mortality studies make it difficult to determine the extent of the increased risk for cardiovascular disease mortality among firefighters. The healthy worker effect makes a clear interpretation of study findings and the comparison of different studies quite difficult.

Nevertheless, several mortality studies have found an increased risk of cardiovascular disease mortality among firefighters despite the expectation that such mortality would be far less than that of the general population due to the physical and health selection factors for entry into the fire service. Studies in Toronto, Hawaii, Connecticut, and New Jersey have found significant increases in cardiovascular disease mortality[4,11,26,36] while several other studies have found cardiovascular disease mortality risks closer to that of the general population.[13,14,27] Only a few studies have found a significant decrease, which would be the expected finding.[29,42]

To date, none of the mortality studies of firefighters have been able to evaluate particular exposures possibly related to cardiovascular disease as part of the data analysis. Some have evaluated age and job duration, but these comparisons have often been limited by small numbers of firefighters in particular age or work experience groups. A few have attempted comparisons with other groups (usually police), but these have also been limited by small numbers and the possibility of different selective factors in the two jobs.

Nor have any of the mortality studies adequately evaluated other cardiovascular disease risk factors. However, a cross-sectional study of a group of Los Angeles firefighters that included the evaluation of other risk factors did find some indication of increased heart disease risk among firefighters while a small prospective study in Boston found no increase.[3,7] Both of these studies have significant limitations, and neither has been repeated.

Given the difficulties with the medical literature, one must evaluate factors in the firefighters' environment that may contribute to cardiovascular disease risk. Two factors stand out. First, firefighters are exposed to high levels of carbon monoxide (a known cardiotoxic substance) at the fire scene, and significant exposure may continue during some aspects of fire suppression, such as when firefighters often do not use SCBAs due to fatigue and the limited air supply of most SCBAs. In addition to the high levels of carbon monoxide, significant exposures may occur to particulates, acrolein, and other respiratory irritants, leading to significant short-term decrements in respiratory function placing further stress on the cardiovascular system.

Secondly, these exposures occur for people working in a job with tremendous physical demands. Firefighters often work at near-maximal cardiac rates for prolonged periods and are subject to high heat loads in addition to the physical loads from wearing SCBAs and other fire protection equipment. These cardiovascular physical demands combined with the possible deleterious effects from carbon monoxide and other fire contaminants undoubtedly place the firefighter at increased risk for acute cardiovascular disease events. The contribution of preexisting coronary heart disease or other risk factors is difficult to determine. While the recent increased use of protective equipment may have reduced the exposure to fire contaminants, this equipment has increased the physical stress on firefighters.

PREVENTION

A number of steps can be taken to reduce the risk of firefighters developing cardiovascular disease.

Most importantly, exposures to carbon monoxide and other fire contaminants must be controlled through proper use of breathing apparatus and proper management of the fire scene. Lighter breathing apparatus and protective equipment must be developed along with SCBAs with longer duration of use. This would reduce the physical stress on the firefighter and eliminate the need to conduct some fire ground activities without the use of SCBAs.

Prevention at the fire scene also includes the presence of adequate staffing to ensure proper rest breaks for fatigued firefighters and adequate personnel for physically demanding operations. Appropriate medical care, including ensuring proper fluid replacement, must be provided at the fire scene to recognize and care for firefighters at risk of heat stroke or overexertion. Fire departments should have medical programs that include appropriate screening for cardiovascular disease risk factors and other health problems as well as programs to help firefighters maintain good physical fitness.

In summary, cardiovascular disease continues to be a significant risk for firefighters. Appropriate preventive and medical programs are needed to help control this problem.

REFERENCES

1. Barnard RJ, Duncan HW: Heart rate and ECG response of fire fighters. J Occup Med 17:247–250, 1975.
2. Barnard RJ, Gardner GW, Diaco NV: "Ischemic" heart disease in fire fighters with normal coronary arteries. J Occup Med 18:818–829, 1976.
3. Barnard RJ, Gardner GW, Diaco NV, Kattus AA: Near-maximal ECG stress testing and coronary artery disease risk factors analysis in Los Angeles City fire fighters. J Occup Med 17:693–695, 1978.
4. Bates JT: Coronary artery disease in the Toronto Fire Department. J Occup Med 29:132–135, 1987.
5. Brandt-Rauf PW, Fallon LF, Tartini T, et al: Health hazards of fire fighters: Exposure assessment. Br J Ind Med 45:606–612, 1988.
6. Davis PO, Dotson CO, Santa Marie DL: Relationship between simulated fire fighting tasks and physical performance measures. Med Sci Sports Exercise 14:65–71, 1982.
7. Dibbs E, Thomas HE, Weiss ST, Sparrow D: Fire fighting and coronary heart disease. Circulation 65:943–946.
8. Duncan HW, Gardner GW, Barnard RJ: Physiological responses of men working in fire fighting equipment in heat. Ergonomics 22:521–527, 1979.
9. Dutton LM, Smolensky MH, Leach CS, et al: Stress levels of ambulance paramedics and firefighters. J Occup Med 20:111–115, 1978.
10. Feuer E, Rosenman K: Mortality in police and firefighters in New Jersey. Am J Ind Med 9:517–527, 1986.
11. Grimes G, Hirsch D, Borgeson D: Risk of death among Honolulu fire fighters. Hawaii Med J 50:82–85, 1991.
12. Guidotti TL: Human Factors in fire fighting: Ergonomic-, cardiopulmonary-, and psychogenic stress-related issues. Int Arch Occup Environ Health 64:1–12, 1992.
13. Guidotti TL: Mortality of urban firefighters in Alberta, 1927–1987. Am J Ind Med 23:921–940, 1993.
14. Hansen ES: A cohort study on the mortality of firefighters. Br J Ind Med 47:805–809, 1990.
15. Hurley BH, Glasser SP, Phelps CP: Cardiovascular and sympathetic reactions to in-flight emergencies among base fire fighters. Aviat Space Environ Med 51:788–792, 1980.
16. International Association of Fire Fighters: Recognition and Prevention of Occupational Heart Diseases, Washington, DC, International Association of Fire Fighters, 1994.
17. Jankovic J, Jones W, Burkhart J, Noonan G: Environmental Study of Fire Fighters. Ann Occup Hyg 35:581–602, 1991.
18. Kalimo R, Lehtonen A, Daleva M, Korinka I: Psychological and biochemical strain in firemen's work. Scand J Work Environ Health 6:179–187, 1980.
19. Kilbom A: Physical work capacity of firemen with special reference to demands during fire fighting. Scand J Work Environ Health 6:48–57, 1980.
20. L'Abbe KA, Tomlinson GA: Scientific Report: Mortality Study of Fire Fighters in Metropolitan Toronto. Industrial Disease Standards Report, Toronto, Ontario, 1992.
21. [Reference deleted.]
22. Lemon PW, Hermiston RT: The human energy cost of fire fighting. J Occup Med 19:558–562, 1977.
23. Levine M, Radford E: Occupational exposures to cyanide in Baltimore fire fighters. J Occup Med 20:53–56, 1978.
24. Louhevaara V, Smolander J, Toumi T, et al: Effects of an SCBA on breathing patterns, gas exchange, and heart rate during exercise. J Occup Med 27:213–216, 1985.
25. Manning JE, Griggs TR: Heart rate in fire fighters using light and heavy breathing equipment: Simulated near maximal exertion in response to multiple work load conditions. J Occup Med 25:215–218, 1983.
26. Mastromatteo E. Mortality in city firemen: A study of mortality in firemen of a city fire department. Arch Ind Health 20:277–283, 1959.
27. Milham S: Occupational mortality in Washington state, 1950–1971. Cincinnati, National Institute for Occupational Safety and Health, 1976.
28. Monson RR Observations on the healthy worker effect. J Occup Med 28:425–433, 1986.
29. Musk AW, Monson RR, Peters JM, Peters RD: Mortality among Boston firefighters, 1915–1975. Br J Ind Med 35:104–108, 1978.
30. Orris P, Kahn G, Melius J: Mortality study of Chicago Firefighters [abstract]. Eighth International Symposium: Epidemiology in Occupational Health. Paris, September 10–12, 1992.
31. Radford E, Levine M: Occupational exposure to carbon monoxide in Baltimore fire fighters. J Occup Med 18:628–632, 1976.
32. Raven PB, Davis TO, Schafer CL, Linnebaur AC: Maximal stress test performance while wearing self-contained breathing apparatus. J Occup Med 19:802–806, 1977.
33. Reich NE: Firefighting and heart disease. Chest 24:304–308, 1953.
34. Romet TT, Frim J. Physiological responses to fire fighting activities, Eur J Appl Physiol 56:633–638, 1987.
35. Sammons J, Coleman R: Fire fighter's occupational exposure to carbon monoxide. J Occup Med 16:543–546, 1974.

36. Sardinas A, Miller JW, Hansen H: Ischemic heart disease mortality of firemen and policemen. Am J Public Health 76:1140–1141, 1986.
37. Sheppard D, Distefano S, Morse L, Becker C: Acute effects of routine firefighting on lung function. Am J Ind Med 9:333–340, 1986.
38. Skoldstrom B. Physiological responses of firefighters to workload and thermal stress. Ergonomics 30:1589–1597, 1987.
39. Sterling TD, Weinkham JJ: Extent, persistence, and constancy of the healthy worker or healthy person effect by all and selected causes of death. J Occup Med 28:348–353, 1986.
40. Stern FB, Halperin WE, Hornung RW, et al: Heart disease mortality among bridge and tunnel officers exposed to carbon monoxide. Am J Epidemiol 128:1276–1288, 1988.
41. Stewart RD, Stewart RC, Stamm W: Rapid estimation of carboxyhemoglobin levels in fire fighters. JAMA 235:390–392, 1976.
42. Takano T, Maeda H: Exposure of fire fighters to carbon monoxide. J Combustion Toxicol 8:89–95, 1981.
43. Tornling G, Gustavson P, Hogstedt C: Mortality and cancer incidence in Stockhom fire fighters. Am J Ind Med 25:219–228, 1994.
44. Treitman R, Burgess WA, Gold A: Air contaminants encountered by fire fighters. Am Ind Hyg Assoc J 41:796–802, 1980.
45. White MK, Hodous TK: Reduced work tolerance associated with wearing protective clothing and respirators. Am Ind Hyg Assoc J 48:304–310, 1987.

MELISSA A. MCDIARMID, MD, MPH
JACQUELINE AGNEW, RN, PHD

REPRODUCTIVE HAZARDS AND FIREFIGHTERS

From the Department of
 Environmental Health Sciences
Division of Occupational Health
The Johns Hopkins School of
 Hygiene and Public Health
Baltimore, Maryland

Reprint requests to:
Melissa A. McDiarmid, MD, MPH
Director, Office of Occupational
 Medicine
U.S. Department of Labor/OSHA
Room N3506
200 Constitution Ave., NW
Washington, DC 20210

The firefighting environment comprises a complex set of hazards, including virtually every known hazard class. Beyond the challenges of characterizing the chemical hazards of even the common structural fire, the post World War II use of plastics and other synthetic materials in construction and for wall and floor covering has greatly multiplied the types of chemical combustion products firefighters encounter. Add to this the physical hazards of heat, physical exertion and noise, the biologic hazards of infectious diseases such as hepatitis B, the human immunodeficiency virus, and, more recently, tuberculosis encountered in emergency medical service runs and the psychosocial stresses of emergency response and shiftwork. This constellation of hazards presents a difficult matrix in which safety and health professionals must prescribe interventions to protect firefighters and is complicated by the additional reality of a nonfixed workplace.

The health effects attributed to exposure to fire smoke have included excesses of nonmalignant respiratory disease in firefighters.[17,74] Less consistent excesses in coronary artery disease have been reported.[6,7,56] Although some results are conflicting, apparent cancer excesses of the gastrointestinal, hematopoietic/lymphatic, and nervous system of firefighters have been reported.[17,51,78]

Little attention has been given to the reproductive hazards of the fire environment. With the exception of a recent report of birth defect excesses in offspring of firefighters[61] and a report describing pregnancy history among women firefighters,[2] no epidemiologic data exist to document reproductive health endpoints in firefighters.

The lack of data on worker reproductive health is not limited to the fire service. The study of workplace reproductive hazards has in many occupational settings been hampered for several reasons, including unique challenges of the various settings. Unlike other physiologic functions, reproductive function is expressed intermittently rather than continuously.[42,43] Therefore, assessment of toxicity after exposure may depend on the timing during which an exposure took place. If the exposure was during a vulnerable period, an adverse outcome may be seen. Otherwise, no apparent harm may be detected. Another complication of reproductive toxicology is the differential species effects that are observed from toxic exposure. Also, reproductive health assessment requires evaluation of a couple or breeding pair attempting pregnancy, as opposed to functions that can be assessed in individuals. These physiologic differences imply the need for alternative approaches to studying reproductive hazards and underscore the necessity of evaluating both members of a couple in determining reproductive health harm.

Until recently, the limited study of occupational reproductive hazards has focused primarily on female reproductive health. This focus has probably been stimulated by the entrance of women into traditional male sectors of the workforce. Although there are gender-mediated differences in chemically induced adverse reproductive outcomes, the majority of well-tested chemicals have demonstrated adverse reproductive outcomes in both men and women.[63] In fact, because of the accessibility of animal and human male gonads and gametes, more agents have been studied and shown to be toxic to male reproductive processes than to female reproductive processes.[44]

MECHANISMS OF REPRODUCTIVE TOXICITY

Adverse effects caused by reproductive toxicant exposure may be manifested at many sites in the complex pathway of reproductive function, beginning with gametogeneses and continuing through gamete interation (fertilization), embryonic and fetal development and growth, parturition, and sexual maturation of the offspring.

The effects observed and the site of insult along this developmental continuum are modulated both by common toxicologic events considered in any xenobiotic exposure and unique aspects of reproductive toxicity such as exposure timing during a vulnerable period of development.

Toxic effects of xenobiotic exposure are classically considered to be a function of an exposure-effect pathway, including systemic absorption, distribution, metabolism, and clearance (excretion), as some critical cellular or subcellular interaction takes place within the target organ to alter normal reproductive function. Anywhere along this exposure-effect continuum, detoxification steps may also alter the toxicity that is ultimately observed. Additionally, repair may ensue subsequent to insult, modifying or completely reversing an effect.

Reproductive toxicants may be broadly classified as acting directly or indirectly.[45] The agents that act indirectly may require metabolic activation before exerting toxicity, a notion reminiscent of the direct/indirect acting carcinogen classification. Alternatively, indirect-acting reproductive toxicants may alter normal reproduction function via metabolism to a direct-acting toxicant or by influencing an enzyme function such as induction or modulation of other enzymatically controlled homeostatic mechanisms.

Agents that act directly may function in one of two ways, the first being via structural similarity to another biologically active molecule. The best examples are oral contraceptives, which limit preovulatory gonadotropin excursions. Several occupational exposures to estrogenic compounds resulting in menstrual abnormalities have been reported.[24,62]

A second mechanism of direct-acting toxicity is that of chemical reactivity. Some alkylating antineoplastic drugs are commonly cited examples of this type of reproductive toxicant. These genotoxic compounds, capable of covalently binding with cellular macromolecules, are mutagenic and many are human carcinogens or teratogenic.[28]

The special case of mutagens must be kept in mind in reproductive hazard identification. Although reproductive toxicologic data on specific toxicants often are not available, mutagenicity data often are. Certainly, a well-characterized mutagen should be considered a potential reproductive toxicant because of its genotoxic nature, even in the absence of reproductive toxicity data. In attempting to bridge the connection between mutagen exposure and reproductive outcome, one recent study showed a statistically significant difference between chromosomal aberrations in dysfertile persons with mutagen exposure compared to dysfertile persons with no mutagen exposure.[30]

Reproductive toxicants are generally detoxified as any xenobiotics, via classical phase 1 and phase 2 metabolic enzyme systems. Nonpolar compounds are usually metabolized by mono-oxygenases to more polar water-soluable compounds before conjugation steps.

Highly reactive compounds such as alkylating agents may be conjugated, sometimes through an epoxide intermediate. The presence of these detoxification enzyme systems has been documented in both ovaries and testes.[25,43–44]

Repair mechanisms also may be activated when detoxification systems are saturated or impaired. Simple repair mechanisms may include enhanced synthesis of biologically important macromolecules. Alternatively, the DNA repair mechanisms that function in genotoxic insult and are more commonly considered in carcinogenic exposures will be important for reproductive toxicants when the insult is genotoxically mediated. While not well characterized, limited evidence documents DNA repair capability in the ovulated oocyte[64] and developing sperm.[14]

Male-Mediated Effects

The biologic plausibility of male-mediated reproductive effects has been increasingly considered, and scientific evidence for such effects has grown rapidly. Wyrobek has reviewed the evidence for male-mediated effects manifested beyond fertilization and the multi-generational context in which reproductive health must be studied.[83]

Spermatogenesis, characterized by rapid cell development in the testes, is a likely target of mutagens that ordinarily interact with dividing cells. Multiple outcomes could result from such interactions, including male infertility and spontaneous abortion. In addition to genotoxic mechanisms, other epigenetic and nongenetic mechanisms modulate male reproductive health at the level of normal physiologic function such as the control of erection and ejaculation. Thus, neurotoxic agents such as lead[31] and inorganic mercury[80] may affect sexual function.

Other effectors of sperm production and male sexual performance include anatomic abnormalities such as cryptorchidism and varicocele, infectious agents such as the mumps virus, host factors such as autoimmunity, and high fever.[80] Environmental agents purported to affect testicular function include alcohol consumption; and cigarette smoke has been reported to cause sperm abnormalities.[83]

Extensively studied pharmacologic agents also have been evaluated for, or observed to cause, reproductive health effects. Detailed studies required in the drug-use approval process, as well as observational studies of therapeutically treated patients, combine to provide these data. Three classes of drugs have been shown to potentially cause some type of male reproductive health affects: hormones affecting secondary sex characteristics, sexual function, and infertility (estrogens, progesterones, testosterone, prednisone); alkylating anticancer drugs causing testicular toxicity and infertility (cyclophosphamide, chlorambucil); and anesthetic gases causing infertility and possibly increased spontaneous abortions (N_2O, halogenated agents).[46]

Occupational studies have reliably demonstrated the often irreversible testicular toxicity of the herbicide dibromochloropropane (DBCP).[80] Other toxicants, especially heavy metals and neurotoxicants, also are being investigated, with some positive evidence of lead causing sperm abnormalities at what previously were considered low concentrations[31] and playing a role in paternally mediated teratogenicity.

A male contribution to spontaneous abortion can be hypothesized via a mutagenic insult to the sperm,[83] paraoccupational exposure resulting in home contamination and maternal exposure,[48] concentration of the agent in semen,[75] and direct transmission of the agent on sperm.[84]

INDUSTRIAL HYGIENE IN THE FIREFIGHTING ENVIRONMENT

A complex variety of fire smoke constituents results from combustion of building materials, furnishings, automobiles, industrial facilities, and their contents. Characterization of potential exposures, including chemicals produced and their concentrations, fuel sources, temperatures, duration of combustion, rate of temperature rise, and fire suppression efforts.[76] In attempting to categorize combustion product chemicals, one group of investigators recently reviewed existing industrial hygiene data on fire smoke. While scores of toxicants have been qualitatively verified in working fires, the authors focused their review on toxicants that were identified quantitatively through personal sampling in working fires versus data derived from test burns and simulations.[47]

The 17 agents identified using the above criteria are displayed in Table 1 with their measured concentration range and, for comparison, as a ratio of the permissible exposure limit (PEL) set by the Occupational Safety and Health Administration (OSHA). This list of toxicants is biased because the toxicants sampled for in monitoring studies are determined by the availability of reliable sampling and analysis methods. Nonetheless, the list represents the best assessment of quantitatively determined toxicants present in working fires.

Reproductive Effects of Combustion Products

Eight of the toxicants in Table 1 exhibited concentration ranges in excess of five times the OSHA permissible exposure limit. A review of the literature of these toxicants

TABLE 1. Concentration Range of Firefighters' Chemical Exposures

Chemical	Exposure Range	PEL Ratio
Acetaldehyde	1–15 ppm	<1 ×
Acrolein	0–98 ppm	0–980 ×
Benzene	0–250 ppm	0–250 ×
Carbon Dioxide	up to 50,000 ppm	up to 10 ×
Carbon Monoxide	0–15,000 ppm	0–425 ×
Chloroform	1–2 ppm	<1 ×
Dichlorofluromethane	1–12 ppm	<1–1.2 ×
Formaldehyde	1–8 ppm	1–8 ×
Hydrogen Chloride	0–280 ppm	0–56 ×
Hydrogen Cyanide	0–75 ppm	0–7.5 ×
Methylene Chloride	up to 0.5 ppm	<1 ×
Nitrogen Dioxide	0–8 ppm	0–1.4 ×
Perchloroethylene	up to 0.2 ppm	<1 ×
Sulfur Dioxide	0–42 ppm	0–21 ×
Toluene	up to 0.275 ppm	<1 ×
Trichloroethylene	up to 0.2 ppm	<1 ×
Trichlorophenol	up to 0.21 ppm	<1 ×

PEL ratio indicates the FF exposure as a function of the OSHA PEL (i.e. 7.5 × = 7.5 times higher than the PEL).

(Table 2) reveals significant animal data documenting reproductive toxicity, mutagenicity, and, less commonly, carcinogenicity evidence for many of these toxicants. Human data are not available concerning the reproductive toxicity of most of these agents.

Several aldehydes are well represented in the literature regarding combustion. Acrolein, a three-carbon aldehyde, is highly mutagenic in a number of in vitro systems and yields positive results in the sister chromatid exchange (SCE) assay using human lymphocytes, indicating its ability to cause chromosomal point mutations in a human cell line.[82]

Formaldehyde, the most extensively characterized aldehyde in the fire environment, has exhibited multiple adverse reproductive effects in animals. There is also evidence of male-mediated effects on fertility in several animal species.[13,54] Formaldehyde is mutagenic in many human test systems.[41,53]

TABLE 2. Reproductive Toxicities of Selected Chemicals in the Fire Environment

Substance	Animal Toxicity		Mutagenicity/ Carcinogencity		Reference
	Male	Female	M	C	
Acrolein		X	H,A		Slott and Hales, 1985 Wilmer et al., 1986 Chung et al., 1984
Benzene	X	X	H,A	H,A	Tatrai et al., 1980 Nawrot and Steples Keller and Synder, 1988 Barlow and Sullivan, 1982 Dowty el al., 1976
Carbon dioxide	X	X			Haring, 1960 Weaver and Scott, 1984 Grote, 1965
Carbon monoxide		X	A		Longo, 1977 Astrup et al., 1972 Kwak et al., 1986
Formaldehyde	X	X	A	A	Yasumura et al., 1983 Davidkova and Basmadzhieva, 1979 Selye and Friedman, 1941 Snyder and Matheson, 1982 Miretskaya and Shavartsma, 1982
Hydrogen chloride		X	A		Stumm-Tegethoff, 1969 Pavlova et al., 1977
Hydrogen cyanide		X			Doherty et al., 1982
Sulfur dioxide	X	X	H,A	A	Mamatsahvili, 1970 Sikov et al., 1978 Gunnison et al., 1987

Adapted from McDiarmid MA, Lees P, Agnew J (et al): Reproductive hazards of firefighting. II. Chemical hazards. Am J Ind Med 19:447-472, 1991.
H, human data; A, animal data; ?, evidence is equivocol; M, mutagenic; C, carcinogenic.

Benzene, the undisputed human leukemogen, has extensive reproductive effects in animals. Adequate studies are not available, but there is one report of menstrual abnormalities and chromosomal insult in humans exposed at high concentrations (>15 ppm).[5] Benzene has been demonstrated to cross the human placenta, allowing it to potentially function as a transplacental carcinogen.

Carbon dioxide, a simple asphyxiant, acts by displacing oxygen and has manifested a number of adverse reproductive outcomes at the level of spermatogenesis, infertility,[55] and developmental abnormalities.[23,79]

Carbon monoxide, a chemical asphyxiant, occupies the oxygen-carrying sites of hemoglobin and has been documented commonly in fires, often at high concentrations. The sensitivity of the human fetus to carbon monoxide has been studied well. The fetus is relatively more sensitive to effects than adults for several physiologic reasons: the fetal carboxyhemoglobin (COHb) concentration is 10–15% greater than the corresponding maternal COHb concentration under steady state conditions; the partial pressure of O_2 in fetal blood is lower, about 20–30 mm Hg, compared with adult levels of 100 mm Hg[36]; the fetal hemoglobin dissociation curve lies to the left of the adult curve, rendering a lower tissue oxygen concentration at identical COHb levels; and there is evidence that the halflife of CO elimination from the fetus is greater than that of the mother.[40] This suggests that maternal exposure to high concentrations of CO during firefighting could be differentially toxic to a developing fetus and should be considered in making policy decisions regarding alternate duty during pregnancy.

Cigarette-smoking mothers and their fetuses have been found to have elevated COHb levels at the time of delivery[37] and have been studied as surrogates for CO-exposed populations. They have an excess of low birthweight births and spontaneous abortions as well as adverse fetal effects.[38] While a number of toxic constituents are found in cigarette smoke, CO is thought to be responsible for these effects. Peak CO concentrations measured in fires exceed 15,000 ppm,[39] but firefighters using self-contained breathing apparatus (SCBA) are not exposed to these concentrations. However, the physiologic difference between adults and fetuses—with the greater relative sensitivity of fetus to CO exposure—bears consideration in policy decisions regarding alternative duties.

Although hydrogen chloride is commonly measured in different fire environments, few reproductive effects have been reported in animals and none in humans. Hydrogen cyanide, which blocks cellular respiration and energy production, is also regularly documented in fire environments but has not been extensively studied for reproductive effects. Sulfur dioxide has been somewhat more extensively studied, yielding positive results in reproductive and developmental studies in animals and male reproductive effects[22] and positive mutagenicity results in several human test systems.[72]

A recent Dutch study documented PAH exposure and uptake by measuring urinary 1-OH-pyrene in firefighting instructors despite relatively brief exposures and the use of respiratory protection.[20]

In addition to the toxicants quantitatively measured in the study of the firefighting environment,[47] a host of toxicants were qualitatively identified, including chemicals such as ammonia and hydrofluoric acid, heavy metals such as cadmium and lead, and polycyclic aromatic hydrocarbons (PAH) such as benzo(a)pyrene, anthracene, and chrysene. Also identified were dioxins, furans, polychlorinated biphenyls (PCBs), and vinyl chloride monomer. There are data documenting reproductive toxicity in animals in many of these agents and some evidence for toxicity to humans from cadmium,[29,68] lead,[69] and vinyl chloride monomer.[5,27]

In the absence of studies specifically examining the reproductive health of firefighters, information may be gained by examining data from working populations exposed to some of the same constituents of fire smoke. A recent epidemiologic review of paternal exposure and spontaneous abortion[71] may provide some insight into the firefighting experience. The authors reviewed 39 studies performed in the previous 18 years that examined the relation between paternal exposure and spontaneous abortion. They considered comments on the study's quality, including method of data collection and verification, power, response rates, and other methodologic effectors of outcome. They reported outcomes based on several groupings of hazard types, some of which are pertinent to firefighter exposure.

Based on job title or employment sector several studies document an elevated relative risk (RR) of spontaneous abortion in workers exposed to metals. One such study of copper smelter workers (exposure to lead, arsenic, mercury, and cadmium) found a relative risk of 1.5 (95% C.I. = 0.9–2.3).[8] Others, more methodologically robust regarding exposure assessment, have shown good evidence for a link between paternal exposure to heavy metals, particularly mercury[3] and lead.[12,35] Negative studies in mercury-exposed dentists[10] and in workers whose jobs would potentially expose them to lead[35] also exist.

Savitz[71] has reviewed another category of toxicant exposure for its relation to spontaneous abortion: a group combining rubber, plastics, and solvents. Included were agents such as vinyl chloride, toluene, benzene, trichloroethane, and "petroleum refinery products," many of which appear in the lists of toxicants found in the firefighting environment. In 1989, Taskinen and colleagues performed a particularly careful study of exposure classification and found an RR of 2.3 for exposure to organic solvents in general and an RR of 1.5 for toluene exposure.[75a] An association with gasoline or benzene exposure in petroleum refineries reported an RR of 2.2 and for trichorethane and methylene chloride exposure an RR of 1.8.[34] A 1976 study of spouses of men exposed to vinyl chloride found an RR of 1.8, with an enhanced effect among younger fathers (RR = 3.7).[27] Although Savitz[71] reported that several studies failing to identify excesses in dry cleaning or rubber workers[49] or those working with other solvents[34] were stronger methodologically, he commented that the weaknesses of many of the negative studies did not exonerate the toxicants. Savitz also reviewed exposures to hydrocarbons and exhausts and found generally null results with the exception of Lindbohm's finding of an RR of 1.4 for chimney sweeps and 1.5 for refinery workers.[34]

There are few data on the potential impact of particulate exposure on reproductive health. However, one in vitro study of human sperm motility when exposed to diesel particle extracts showed moderate but progressively stronger effects on motility with duration of exposure and increased dose.[19]

While the epidemiologic evidence for an adverse effect of work-related exposure to chemical toxicants on reproductive health is not consistently firm, the number and quality of studies is increasing with suggestively positive results. Taken with animal data and biologically plausible mechanisms of reproductive toxicity, mounting evidence suggests that many suspect reproductive toxicants are likely to be toxic in humans and deserve respect and treatment as such while confirmatory studies are performed.

NONCHEMICAL HAZARDS

A number of nonchemical hazards within the firefighting environment may affect the reproductive health of firefighters. These hazards have been reviewed by Agnew et al.[1] and are summarized in Table 3.

Heat

A hazard deserving specific discussion is heat. In addition to the extreme of environmental heat firefighters encounter, wearing turnout gear or encapsulated suits for

TABLE 3. Summary of Potential Reproductive Effects of Non-Chemical Exposures

Agent	Animals		Human		Reference
	Male	Female	Male	Female	
Hyperthermia	Decreased sperm number	Fetal malformations	Decreased sperm number	Birth defects, with maternal fever	Henderson et al., 1986 Edwards et al., 1986 Pleet et al., 1981 Clarren et al., 1979 Procope, 1965
			Abnormal sperm Delayed conception	Hearing loss in children of exposed mothers(?)*	Lalande et al, 1986 Rachootin and Olsen, 1983
Physical activity		Few effects noted on fetus	Trauma: testicular damage, hormonal change, impotence	Amenorrhea, strenuous job: prematurity and low birth weight	Lotgering et al., 1985 Steeno and Pangkahila, 1984 Armstrong, 1986 Warren, 1983 Naeye and Peters, 1982
				Heavy lifting or standing on job: miscarriages(?)*	Mamelle et al, 1984 Saurel-Cubizoles et al., 1987 Saurel-Cubizoles and Kaminski, 1987
				A 20+ weeks pregnant: problem with balance and agility	Taskinen et al., 1986 McDonald et al., 1988 AMA Council Sci Affairs, 1984
Noise		Increased litter resorption and fetal mortality, decreased fetal weight		Increased rates of birth defects and low birth weight(?)*	Kimmel et al., 1976 Nawrot et al., 1980 Cook et al., 1982 Edmonds et al., 1979 Jones and Tauscher, 1978
				Hormonal disturbances	Knipchild et al., 1981 Schell, 1981
		Fetal malformations(?)*		Idiopathic infertility	Rachootin and Olsen, 1983
Psychological stress	Decreased testosterone one level		Decreased testosterone levels	Amenorrhea	McGrady, 1984 U.S. Congress, OTA 1985 Fries et al., 1974
			Negative behavioral effects	Negative behavioral effects	

* ? indicates that strength of study design or results do not justify definite conclusions.
Adapted from Agnew J, McDiarmid MA, Lees PSJ, Duffy R: Reproductive hazards of fire fighting 1. Non-chemical hazards. Am J Ind Med 19:433–445, 1991.

special duties also may raise the firefighter's core temperature. Heat causes well-documented insult to the spermatogenic process.[26] The number of human sperm decline and morphology is altered with increases in ambient temperature.[50,67] Although this effect is apparently reversible, the time to normal sperm production is a function of degree and duration of hyperthermia.

The effects of heat on the reproductive health of firefighters have not been studied, but several other heat-exposed working cohorts have been examined for reproductive effects. One Danish case-control study found that men with sperm abnormalities reported exposure to heat more commonly than fertile controls (odds ratio = 1.8; 95% confidence interval = 1.2, 2.6). The study also compared couples who initially experienced a conception delay with couples who successfully conceived within 1 year. An OR of 1.8 (95% CI = 1.4, 2.4) was found for exposure of the father to heat, and the effect was sustained after adjustment for maternal effectors. Interestingly, exposure of the mother to heat also was statistically related to conception delay, OR = 1.5 (95% CI = 1.0, 2.4), and approached significance after adjustment for other confounders, OR = 1.4 (95% CI = 0.9–2.3)[68]

More recently, Italian investigators studied the effect of exposure to the heat of ceramic ovens on fertility and semen quality.[18] The primary area in which workers spent time often registered a wet bulb globe temperature of 38°C. In a self-reported interview, 7.6% of exposed workers and 1.1% of controls reported being in a childless marriage (p = 0.05), and 23.1% of exposed workers and 12.5% of controls complained of difficulty conceiving (p = 0.05). Time until pregnancy was longer in heat-exposed workers than nonexposed controls: 3.24 versus 0.41 months, p = 0.05) for the most recent pregnancy and 3.33 versus 1.27 months (p = 0.05) for all pregnancies.

The effect of heat on the outcome of pregnancy in women firefighters is controversial. In animals, core temperature elevations of 1.5–2.5°C may be teratogenic.[15] Studies of women who are febrile during pregnancy have raised questions about associated congenital abnormalities of offspring.[11,66] Studies of hot tub and sauna use by pregnant women have suggested an association with neural defects.[52] One professional organization has considered a limit of 38.9°C as a minimum core temperature likely to pose a teratogenic hazard to human embryos or fetuses.[4]

While some investigators have concluded that hot working environments pose no threat to pregnancy outcome generally,[4] or to women firefighters specifically,[16] care must be taken before dismissing the hazard of hyperthermia.

A study characterizing core temperature excursions in firefighters in turnout gear completing standard firefighting activities in hot ambient environments of 38.1–41.2°C documented core temperatures of 38.5–39°C.[77] The authors note that a firefighter might spend up to 30 minutes in ambient temperatures of 60–300°C in a structural fire, suggesting that the core temperatures recorded in the simulation underestimate the actual firefighting experience. Studies of the effect of turnout gear, other protective clothing, and respirator use during exercise have documented core temperatures in male firefighters ranging 37.9°–39°C.[73,81] The impact of heat, protective clothing, and work of breathing through a respirator on already heat-intolerant pregnant women is unknown. However, given the American Medical Association's core temperature threshold of 38.9° for teratogenesis, firefighting may pose a risk to the fetus.

That firefighting requires high levels of physical activity is undisputed and has been studied well.[21,32–33,70] The effects of physical activity on the hypothalamic-pituitary-gonadal axis are well documented but not studied specifically in firefighters (Table 3).

Noise exposure is well known among firefighters, with hearing loss a documented problem for this occupation.[59] The decibel exposure to noise in the firefighting environment is in the 60–85 dBA range[57,60] below the OSHA standard of 90 dBA for an 8-hour TWA exposure. Simulated response calls average mid-90 dBA range, with excursions up to 118 dBA.[58] There is spotty evidence for noise affecting reproductive outcome, as summarized in Table 3. The Danish study discussed above also inquired about effects of occupational noise exposure.[68] The authors reported that subfertile noise-exposed women were more likely to experience hormonal disturbances (OR = 2.2, 95% CI = 1.5–3.1) and idiopathic infertility (OR = 2.4, 95% CI = 1.4–4.0) than nonexposed fertile controls.

Also in the Danish study, in a cohort of couples who first experienced a delay in conception of greater than 1 year, both men (OR = 1.3, 95% CI = 1.0–1.6) and women (OR = 1.7, 95% CI = 1.3–2.3) were more likely to have been in noise-exposed jobs than couples who conceived a healthy child in 1 year. These findings recommend further study.

Of the nonchemical hazards of firefighting on reproductive health, it appears that hyperthermia poses the strongest risk for both males and pregnant female firefighters. The risk of other nonchemical hazards in less clear, but some of the positive studies suggest the need for further inquiry.

CONCLUSION

The reproductive risk to firefighters cannot be fully determined because of limitations of both exposure data and toxicity profiles of the toxicants of interest. However, the recognized health risks of firefighting must be enlarged to include the reproductive health of the firefighter. Traditional occupational health strategies to minimize worker exposure to hazardous substances, such as worker education and vigilant use of self-contained breathing apparatus, including during the overhaul phase of firefighting, will also minimize exposure to reproductive toxicants.

This chapter summarizes data documenting three populations at potential risk from a reproductive toxicant: men, women, and developing fetuses. Since adverse reproductive outcomes can be mediated through the male or the female, a reproductive health policy that addresses only pregnant women or women in general is inadequate. In a recent study of female firefighters, most reported that their mates were also firefighters.[2] This suggests an opportunity for a "double dose" of fire-related toxicant exposure to their offspring.

While a comprehensive reproductive health policy must address all three potential targets of a reproductive toxicant, pregnant firefighters deserve special attention. Due to the special sensitivity of human fetuses to carbon monoxide and to the potential for hypoxic episodes, the authors believe that pregnant firefighters should be given alternative duty from the time pregnancy is discovered and that pregnant firefighter candidates temporarily defer training.

Risk reduction through exposure prevention is the principal strategy to maintain health. Classical approaches—including SCBA use, worker education regarding hazards, good work practices, and other administrative controls such as a prudent comprehensive reproductive health policy—taken together, provide the framework to prevent reproductive health harm to both male and female firefighters.

REFERENCES

1. Agnew J, McDiarmid MA, Lees PSJ, Duffy R: Reproductive hazards of fire fighting 1. Non-chemical hazards. Am J Ind Med 19:433–445, 1991.
2. Agnew J, McDiarmid MA, Fitzgerald S: A survey of reproductive health of women firefighters [manuscript].
3. Alcser KH, Brix KA, Fine LJ, et al: Occupational mercury exposure and male reproductive health. Am J Ind Med 15:517–529, 1989.
4. American Medical Association Council on Scientific Affairs: Effects of physical forces on the reproductive cycle. JAMA 251:247-250, 1984.
5. Barlow SM, Sullivan FM: Reproductive Hazards of Industrial Chemicals. New York, Academic Press, 1982.
6. Barnard RJ, Weber JS: Carbon monoxide: A hazard to firefighters. Arch Environ Health 34:255–257, 1978.
7. Bates JT: Coronary artery disease deaths in the Toronto fire department. J Occup Med 29:132–135, 1987.
8. Beckman L, Nordstrom S: Occupational and environmental risks in and around a smelter in northern Sweden. Hereditas 97:1–7, 1982.

9. Bell JU, Thomas JA: Effects of lead on mammalian reproduction. In Singhand RL. Thomas JA (eds): Lead Toxicity. Baltimore, Urban & Schwarzenburg, 1980, pp 169–185.
10. Brodsky JB, Cohen EN, Whitcher C, et al: Occupational exposure to mercury in dentistry and pregnancy outcome. J Am Dent Assoc 11:779–780, 1985.
11. Clarren SK, Smith DW, Harvey MA, et al: Hyperthermia: A prospective evaluation of a possible teratogenic agent in man. J Pediatr 95:81–33, 1979.
12. Cordier S, Deplan F, Mandereau L, Hemon D: Paternal exposure to mercury and spontaneous abortions. Br J Ind Med 48:375–381, 1991.
13. Davidkova R, Basmadzhieva K: Changes in protein and nucleic acid metabolism as one of the methods for evaluating gonadotoxic action. Probl Khig 4:101–109, 1979.
14. Dixon RL, Lee IP: Pharmacokinetic and adaption factors involved in testicular toxicity. Fed Proc 39:66–72, 1980.
15. Edward MJ: Hyperthermia as a teratogen: A review of experimental studies and their clinical significance. Teratog Carcinog Mutagen 6:563–582, 1986.
16. Evanoff BA, Rosenstock L: Reproductive hazards in the workplace: A case study of women firefighters. Am J Ind Med 9:503–515, 1986.
17. Feuer E, Rosenman K: Morality in police and firefighters in New Jersey. Am J Ind Med 9:517–527, 1986.
18. Figa-Talamanca I, Dell 'Orco V, Pupi A, et al: Fertility and semen quality of workers exposed to high temperatures in the ceramic industry. Reprod Toxicol 6:517–523, 1992.
19. Fredricsson B, Moller L, Pousette A, Westerholm R: Human sperm motility is affected by plasticizers and diesel particle extracts. Pharmacol Toxicol 72:128–133, 1993.
20. Fuenekes F, Jongeneelen F, Laan H vd, Schoonhof F: Uptake of polycyclic aromatic hydrocarbons among trainers in a firefighting training facility.
21. Guidotti T: Human factors in firefighting: Ergonomic, cardiopulmonary, and psychogenic stress-related issues. Occup Environ Health 64:1–12, 1992.
22. Gunnison AF, Sellakumur A, Currie D, Snyder EA: Distribution, metabolism, and toxicity of inhaled sulfur dioxide and endogenously generated sulfite in the respiratory tract of normal and sulfite oxidase-deficient rates. J Toxicol Environ Health 21:141, 1987.
23. Haring OM: Cardiac malformation in rats induced by exposure of the mother to carbon dioxide during pregnancy. Circ Res 8:1218–1227, 1960.
24. Harrington JM, Stein GF, Rivera RO, de Morales AV: The occupational hazards of formulating oral contraceptives: A survey of plant employees. Arch Environ Health 33:12–15, 1978.
25. Heinricks WL, Juchan MR: Extraleynatic drug metabolism: The gonads In Gram TE (ed): Extrahepatic Metabolism of Drugs and Other Foreign Compounds. New York, SP Medical and Scientific Books, 1984, pp 319–332.
26. Henderson J, Baker HWG, Hanna PH: Occupation-related male infertility: A review. Clin Reprod Fertil 4:87–106, 1986.
27. Infante PF, Wagoner JK, McMichael AJ, et al: Genetic risks of vinyl chloride. Lancet 1:734, 1976.
28. International Agency for Research on Cancer: Chemicals, industrial processes, and industries associated with cancer in humans. IARC Monogr Suppl 4:292, 1982.
29. International Agency for Research on Cancer: Overall evaluation of carcinogenicity: An updating of IARC monographs volumes 1 to 42. IARC Monogr 7 (Suppl), 1987.
30. Kucerova M, Gregor V, Horacek J, et al: S: Influence of different occupations with possible mutagenic effects on reproduction and level of induced chromosomal aberrations in peripheral blood. Mutat Res 278:19–22, 1992.
31. Lancranjan I, Popescu H, Gavanescu O: Reproductive ability of workmen occupationally exposed to lead. Arch Environ Health 30:396–401, 1975.
32. Lemon PWR, Hermiston RT: Physiological profile of professional fire fighters. J Occup Med 19:337–340, 1977.
33. Lemon PWR, Hermiston RT: The human energy cost of firefighting. J Occup Med 19:558–562, 1977.
34. Lindbohm M-L, Hemminki K, Bonhomme MG, et al: Effects of paternal occupational exposure on spontaneous abortions. Am J Public Health 81:1029–1033, 1991.
35. Lindbohm M-L, Sallmen M, Anttila A, et al: Paternal occupational lead exposure and spontaneous abortions. Scand J Work Environ Health 17:95–103, 1991.
36. Longo LD: Carbon monoxide effects on oxygenation of the fetus in utero. Science 194:523–525, 1976.
37. Longo LD: The biological effects of carbon monoxide on the pregnant women, fetus and newborn infant. Am J Obstet Gynecol 129:69–103, 1977.
38. Longo LD: Environmental pollution and pregnancy: Risks and uncertainties for the fetus and infant. Am J Obstet Glynecol 137:162–173, 1980.
39. Lowry WT, Juarez L, Petty CS, Roberts B: Studies of toxic gas production during actual structural fires in the Dallas area. J Forensic Sci 30:59–72, 1985.

40. Margulies SL: Acute carbon monoxide poisoning during pregnancy. Am J Emerg Med 4:516–519, 1986.
41. Martin CN, McDermid AC, Garner RC: Testing of known carcinogens and noncarcinogens for their ability to induce unscheduled DNA synthesis in HeLa cells. Cancer Res 38:2621–2627, 1978.
42. Mattison DR: Drugs, xenobiotics and the adolescent: Implications for reproduction. In Soyka LF, Redmond GP (ed): Drug Metabolism in the Immature Human. New York, Raven Press, 1981, pp 129–143.
43. Mattison DR, Nightengale MS: Environmental factors in human growth and development. In Hunt VR, Smith MK, Worth D (eds): Prepubertal Ovarian Toxicity. Cold Spring Harbor, NY, Cold Spring Harbor Laboratory, 1982, Banbury Report 11.
44. Mattison DR: The mechanisms of action of reproductive toxins. Am J Ind Med 4:65–79, 1983.
45. Mattison DR, Thomford PJ: The mechanism of action of reproductive toxicants. Toxicol Pathol 17:364–376, 1989.
46. McDiarmid MA: Occupational exposure to pharmaceuticals. In Paul MA (ed): Occupational and Environmental Reproductive Hazards: A Guide for Clinicians. Baltimore, Williams & Wilkins, 1994, pp 280-295.
47. McDiarmid MA, Lees P, Agnew J, et al: Reproductive hazards of firefighting. II. Chemical hazards. Am J Ind Med 19:447–472, 1991.
48. McDiarmid MA, Weaver V: Fowling one's own nest revisited. Am J Ind Med 24:1–9, 1993.
49. McDonald AD, McDonald JC, Armstrong B, et al: Father's occupation and pregnancy outcome. Br J Ind Med 46:329–333, 1989.
50. Mieusset R, Bujan L, Mansar A, et al: Effects of artificial cryptorchidism on sperm morphology. Fertil Steril 47:150–155, 1987.
51. Milham S Jr: Occupationally mortality in Washington State 1950–71. Cincinnati, NIOSH Division of Surveillance, Hazard Evaluation and Field Studies, 1976, NIOSH publication 76–175-C.
52. Miller P, Smith DW, Shepard NH: Small head size after inutero exposure to atomic radiation. Lancet 2:784–787, 1972.
53. Miretskaya LM, Shvartsma PY: Studies of chromosome aberrations in human lymphocytes under the influence of formadehyde. I. Formaldehyde treatment of lymphocytes in vitro. Tsitologiia 24:1056–1060, 1982.
54. Morpurgo G, Bellincampi D, Gualandi G, et al: Analysis of mitotic non-disjunction with Aspergillus nidulans. Environ Health Perspect 31:81–95, 1979.
55. Mukherjee DP, Singh SP: Effect of increased carbon dioxide in inspired air on the morphology of spermatozoa and fertility of mice. J Reprod Fertil 13:165–167, 1967.
56. Musk AW, Monson RR, Peters RK: Mortality among Boston firefighters, 1915–1975. Br J Ind Med 35:104–198, 1978.
57. National Institute for Occupational Safety and Health: Newburgh Fire Department, Newburgh, NY, 1982, health hazard evaluation report HETA 81-059-1045.
58. National Institute for Occupational Safety and Health: The City of New York Fire Department, New York, NY, 1985, health hazard evaluation report HETA 81-459-1603.
59. National Institute for Occupational Safety and Health: International Association of Fire Fighters. Cincinnati, 1988, health hazard evaluation report HETA 86-454-1890.
60. National Institute for Occupational Safety and Health: Memphis Fire Department, Memphis, TN, 1990, health hazard evaluation report HETA 86-138-2017.
61. Olshan AF, Teschke K, Baird PA: Birth defects among offspring of firemen. Am J Epidemiol 131:312–321, 1990.
62. Pacynski A, Budzynska A, Przylecki S: Hiperestrogenizm v pracownikow zakladow farmaceutyczaych i ich dzieci jako choroba zawodowa. Endokrynol Pol (Warsaw) 22:149–154, 1971.
63. Paul M, Himmelstein J: Reproductive hazards in the workplace: What the practitioner needs to know about chemical exposure. Obstet Gynecol 71:921–938, 1988.
64. Perdersen RA, Manigia F: Ultraviolet light induced unscheduled DNA synthesis by resting and growing mouse oocytes. Mutat Res 49:425–429, 1978.
65. Peters JM, Theriault GP, Fine LJ, Wegman DH: Chronic effect of fire fighting on pulmonary function. N Engl J Med 291:1320–1322, 1974.
66. Pleet HG, Graham JM, Smith DW: Central nervous system and facial defects associated with maternal hyperthermia in four to 14 weeks gestation. Pediatrics 67:785–789, 1981.
67. Procope BJ: Effect of repeated increase of body temperature on human sperm cells. Int J Fertil 10:333–339, 1965.
68. Rachootin P, Olsen J: The risk of infertility and delayed conception associated with exposure in the Danish workplace. J Occup Med 25:394–402, 1983.
69. Rom WN: Effects of lead on the female reproduction: A review. Mt Sinai J Med 43:542–552, 1976.
70. Romet TT, Frim J: Physiological responses to fire fighting activities. Eur J Appl Physiol 56:633–638, 1987.
71. Savitz D, Sonnenfeld N, Olshan A: Review of epidemiologic studies of paternal occupational exposure and spontaneous abortion. Am J Ind Med 25:361–383, 1994.

72. Shapiro R: Genetic effects of bisulfite (sulfur dioxide). Mutat Res 39:149–176, 1977.
73. Smolander J, Louhevara V, Kohonen O: Physiological strain in work with gas protective clothing at low ambient temperature. Am Ind Hyg Assoc J 46:720–723, 1985.
74. Sparrow D, Bosse T, Rosner B, Weiss S: The effect of occupational exposure on pulmonary function. A longitudal evaluation of fire fighter and non-fire fighters. Am Rev Respir Dis 125:319–322, 1982.
75. Stachel B, Dougherty TC, Lahl U, et al: Toxic environmental chemicals in human semen: Analytical method and case studies. Andrologia 21:282–291, 1989.
75a. Taskinen H, Antila A, Lindbohm M-L, et al: Spontaneous abortions and congenital malformations among the wives of men occupationally exposed to organic solvents. Scand J Work Environ Health 15:345–352, 1989.
76. Terrill JB, Montogomery RR, Reinhart CF: Toxic gases from fires. science 200:1343–1347, 1978.
77. Veghte JH: Physiologic Response of Fire Fighters Wearing Bunker Clothing (Phase 2). Beaver Creek, Ohio, Biotherm, 1987.
78. Vena JE, Fielder RC: Mortality of a municipal worker cohort: IV: Fire fighters. Am J Ind Med 11:671–684, 1987.
79. Weaver TE, Scott WJ Jr: Acetazolamide teratogenesis: Association of maternal respiratory acidosis and ectrodactyly in C57BL/6J mice. Teratology 30:187–193, 1984.
80. Wharton MD: Adverse reproductive outcomes: The occupational health issues of the 1980s. Am J Public Health 73:15–16, 1983.
81. White MK, Hodus TK: Reduced work tolerance associate with wearing protective clothing and respirators. Am Ind Hyg J 48:304–310, 1987.
82. Wilmer JL, Erexson GL, Kligerman AD: Attenuation of cytogenic damage bu 2-mercaptoerhane-sulfonate in cultured human lymphocytes exposed to cyclophosphamide and its reaction metabolites. Cancer Res 46:203–210, 1986.
83. Wyrobek A: Methods and concepts in detecting abnormal reproductive outcomes of paternal origin. Reprod Toxicol 7:3–16, 1993.
84. Yazigi RA, Odem RR, Polakoski KL: Demonstration of specific binding of cocaine to human spermatozoa. JAMA 266:1956–1959, 1991.

RANDY L. TUBBS, PhD

NOISE AND HEARING LOSS IN FIREFIGHTING

From the Division of Surveillance,
Hazard Evaluations, and Field
  Studies
National Institute for Occupational
  Safety and Health
Cincinnati, Ohio

Reprint requests to:
Randy L. Tubbs, PhD
National Institute for
  Occupational Safety
  and Health
Mail Stop R-11
4676 Columbia Parkway
Cincinnati, OH 45226

Mention of company names or products does not constitute endorsement by the National Institute for Occupational Safety and Health.

Hearing loss from occupational noise exposure is a completely preventable disease. This premise is based on a simple fact: if we are able to stop excessive noise from entering the ear, no loss of hearing can result from an occupational noise source. Even though this statement has been accepted as truth for a long time, workers continue to lose hearing as a result of their occupation, despite regulations and education. One occupation that has received the attention of researchers at the National Institute for Occupational Safety and Health (NIOSH) since 1980 is firefighting. The institute's Health Hazard Evaluation Program received a request to investigate an apparently high degree of hearing loss in a municipal fire department in October 1980.[30] Since this time, several additional fire departments and emergency medical services (EMS) have been surveyed by NIOSH investigators, who measure the occupational noise exposures and the hearing ability of the firefighters and EMS personnel.

Many of the firefighters' noise exposures are obvious, including the sirens and air horns that warn the public of the approaching emergency vehicles, as well as the roar of the diesel engines that power the trucks. As mechanized equipment replaces older hand tools, the equipment used on the fire scene is becoming louder. Even the living quarters of firefighters can be noisy, being located on or near major highways for easy access to roads. The stations are often built with shiny, hard-surfaced materials that reflect the noise around the living quarters. Many kitchens have metal stoves and heavy metal cooking utensils that contribute to the overall noise emissions. Departments that have the additional responsibility of airport fire and rescue also hear the aircraft landings and take-offs.

Many of the tasks firefighters perform depend

on the auditory ability of the person. It is nearly impossible to see in a smoke-filled room. This loss of visual acuity is typified in the training adage of "go for the glow," which directs firefighters to the area of the working fire. They are also taught to listen for moans and cries when conducting a rescue search. In addition, they must listen to and respond to radio communications as well as for the warning sound of an air horn that signals firefighters to leave the interior of a building because of imminent danger. A loss of auditory acuity can literally be a life-and-death situation for people engaged in this type of work. These critical hearing requirements have led to a body of research that has quantified the hearing abilities of firefighters and the amount of noise to which they are exposed in their occupation.

EVALUATION CRITERIA FOR OCCUPATIONAL NOISE EXPOSURE AND AUDIOMETRIC TESTING

Occupational Noise

Noise-induced loss of hearing is an irreversible, sensorineural condition that progresses with exposure. Although hearing ability declines with age (presbycusis) in all populations, exposure to noise produces hearing loss greater than that resulting from the natural aging process. This noise-induced loss is caused by damage to nerve cells of the inner ear (cochlea) and, unlike some conductive hearing disorders, cannot be treated medically.[38] While loss of hearing may result from a single exposure to a brief impulse noise or explosion, such traumatic losses are rare. In most cases, noise-induced hearing loss is insidious. Typically, it begins to develop at 4,000 or 6,000 Hz (the hearing range is 20 to 20,000 Hz) and spreads to lower and higher frequencies. Often, material impairment has occurred before the condition is clearly recognized. Such impairment is usually severe enough to cause permanent damage to a person's ability to hear and understand speech under everyday conditions. Although the primary frequencies of human speech range from 200 to 2,000 Hz, research has shown that the consonant sounds, which enable people to distinguish words such as *fish* from *fist,* have still higher frequency components.[26]

The A-weighted decibel [dB(A)] is the preferred unit for measuring sound levels to assess worker noise exposure. The dB(A) scale is weighted to approximate the sensory response of the human ear to sound frequencies near the threshold of hearing. The decibel unit is dimensionless and represents the logarithmic relationship of the measured sound pressure level to an arbitrary reference sound pressure (20 μPa, the normal threshold of human hearing at a frequency of 1,000 Hz). Decibel units are used because of the large range of sound pressure levels that are audible to the human ear. Because the dB(A) scale is logarithmic, increases of 3 dBA, 10dBA, and 20 dBA represent a doubling, tenfold, and 100-fold increase of sound energy, respectively. It should be noted that noise exposures expressed in decibels cannot be averaged by taking the simple arithmetic mean.

The Occupational Safety and Health Administration (OSHA) standard for occupational exposure to noise (29 CFR 1910.95) specifies a maximum permissible exposure limit (PEL) of 90 dB(A) for a duration of 8 hours per day.[5] The regulation, in calculating the PEL, uses a 5-dB time/intensity trading relationship, or exchange rate. This means that a person may be exposed to noise levels of 95 dB(A) for no more than 4 hours, to 100 dB(A) for 2 hours, and so on. Conversely, up to 16 hours exposure to 85 dB(A) is allowed by this exchange rate. NIOSH, in its Criteria for a Recommended Standard, proposed a recommended exposure limit of 85 dB(A) for 8 hours, 5 dB less than the OSHA standard.[18] The NIOSH 1972 criteria document also used a 5-dB time/intensity trading relationship in calculating exposure limits. However, in 1995, NIOSH changed its official recommendation for an exchange rate from 5 dB to 3 dB.[19] The American Conference of Governmental Industrial Hygienists also changed its threshold limit value (TLV) in 1994 to a more protective 85 dB(A) for an 8-hour exposure, with the stipulation that a 3-dB exchange rate be used to calculate time-varying noise exposures.[2] Thus, a worker can be exposed to 85 dB(A) for 8 hours, but to no more than 88 dB(A) for 4 hours or 91 dB(A) for 2 hours.

The duration and sound level intensities can be combined to calculate a worker's daily noise dose according to the formula:

$$\text{Dose} = 100 \times (C_1/T_1 + C_2/T_2 + \ldots + C_n/T_n)$$

where C_n indicates the total time of exposure at a specific noise level, and T_n indicates the reference duration for that level as given in table G-16a of the OSHA noise regulation.[5] During any 24-hour period, a worker is allowed up to 100% of the daily noise dose. Doses greater than 100% are in excess of the OSHA PEL.

The OSHA regulation has an additional action level of 85 dB(A); an employer shall administer a continuing, effective hearing conservation program when the time-weighted average (TWA) value exceeds the action level. The program must include monitoring, employee notification, observation, audiometric testing, hearing protectors, training, and recordkeeping. All of these requirements are included in 29 CFR 1910.95, paragraphs (*c*) through (*o*).

Finally, the OSHA noise standard states that when workers are exposed to noise levels in excess of the OSHA PEL of 90 dB(A), feasible engineering or administrative controls shall be implemented to reduce the workers' exposure levels. However, in 1983, a compliance memorandum (CPL 2–2.35) directed OSHA compliance officers not to cite employers for lack of engineering controls until workers' TWA levels exceed 100 dB(A), so long as the company has an effective hearing conservation program in place. Even in TWA levels in excess of 100 dB(A), compliance officers are to use their discretion in issuing fines for lack of engineering controls.

Audiometric Tests

Audiometric test results are combined according to different criteria to ascertain the hearing handicap for speech perception or for the determination of noise effects on hearing. A criterion proposed by NIOSH in its criteria document for occupational noise exposure is intended to determine the amount of handicap in speech perception and communication abilities.[18] It averages the hearing level, or amount of hearing deficit, in decibels (dB HL)(ANSI S3.6-1989) at the pure-tone frequencies of 1,000, 2,000, and 3,000 Hz for each ear.[3] The criterion incorporates a 25-dB "low-fence" value. This means that the dB HL average value must exceed 25 dB before a hearing handicap is said to exist. The percentage of handicap is calculated by multiplying each decibel in excess of 25 dB HL by 1.5%. For example, an average dB HL of 40 for this metric would represent a 22.5% hearing handicap.

A second formula has been proposed by the American Academy of Otolaryngology–Head and Neck Surgery.[1] The criterion combines the pure-tone frequencies of 3,000, 4,000, and 6,000 Hz and states that an average change of 20 dB at these frequencies will serve as an otologic referral criterion. The referral is for otologic evaluation to determine the nature and cause of the hearing impairment. This combination is most sensitive to the sensorineural effects on the ear from noise because of the propensity of these frequencies to deteriorate sooner when a person is exposed to loud noises.[13]

Finally, a criterion proposed by Eagles et al. for single-frequency hearing impairment determination also uses a low fence of 25 dB HL.[6] With this criterion, any person who has a hearing level of 26 dB HL or greater at any single frequency is classified as having some degree of hearing loss. The degree of loss could range from mild (26–40 dB HL) to profound (>90 dB HL). This criterion differs from the other two criteria in that it looks at single test frequencies rather than average hearing levels across several frequencies.

OCCUPATIONAL NOISE RESEARCH

The documentation of noise exposures in the fire service was first reported in the literature in the 1970s. Reischl and colleagues reported that California Department of Forestry firefighters were exposed to excessive noise during emergency "code 3" activities, with noise levels approaching 115 dB(A).[22] Reischl et al. used dosimeters that were turned on only during code 3 operations because of the limited information that the dosimeter could provide, i.e., a single cumulative dose percentage. The daily noise doses were calculated mathematically, using the equivalent continuous sound pressure level method.[25] To make the calculations, the firefighter's day was divided into four different categories, and averaged noise levels for each of the categories were plugged into the dose equation. The four categories were code 3 operations, fire scene environment, return travel to station, and station environment. Fire station log-book records were used to estimate times at the station and the number of emergency runs. The calculated noise levels suggest that firefighters were exposed to noise levels that exceeded the OSHA PEL of 90 dB(A). These excessive exposures were found for firefighters occupying nearly all of the riding positions on the vehicles, with the one exception of the tillerman on the truck. In addition to the dosimeter measurements, the researchers also performed octave-band analyses on fire engines and fire trucks during simulated code 3 operations and found high noise exposures to firefighters in all riding positions on the vehicles.

Newburgh Fire Department Survey. NIOSH began its involvement with firefighter noise exposure in 1981 at the Newburgh, New York, Fire Department at the request of the International Association of Fire Fighters.[30] The noise survey for this department used noise dosimeters that are capable of recording 1-minute average noise exposures and storing them for later analysis. This capability allows one to view the noise from an 8-hour work shift for every minute of the measurement period. If one coordinates the noise with the firefighter's activities during the shift, the noise exposure associated with each event can be documented. During simulated emergency runs for each of the department's vehicles, noise levels at each riding position were recorded on audiotape for later octave-band analysis. Additionally, tape recordings were made at the engine's pump panel while drafting water and also at the operator's position during the use of a cutting tool, air compressor, floating pump, and gasoline-powered electric generator.

The tape recordings of the simulated emergency runs captured noise levels up to 116 dB(A) for a 30-second period. The lowest noise exposure measured in one of the vehicle riding positions was 100 dB(A). The noise at the pump panel of the engines and near the various tools and equipment tested at the Newburgh Fire Department was always in excess of 90 dB(A). All of the above measurements were made over a relatively short time, but a firefighter's daily noise exposure must take into account variation in noise exposure over the entire work shift.

The noise dosimeter results for this department revealed a much different noise picture. The 8-hour noise TWAs recorded over two consecutive days ranged from 63 to 85 dB(A), values that are less than the 8-hour OSHA PEL of 90 dB(A). Even when the TWAs are adjusted for longer work shifts of 16 or 24 hours, only 2 of 16 measurements exceeded the evaluation criterion. Of course, the number of emergency responses conducted during the recording period greatly influences the measured worker exposure. There were fewer emergency runs than usual during the two noise survey days at the Newburgh Fire Department.

New York City Fire Department Survey. NIOSH's involvement with firefighter noise exposure continued in 1982 with a noise survey of the large fleet of engine and ladder vehicles of the Fire Department for the City of New York.[31] Fifty-five of the department's 350 engine and ladder trucks were brought to the fire training facility and the chauffeurs' road course to measure the noise levels at the firefighters' riding positions during simulated emergency runs. The course is approximately 1 mile long, with several turns, curves, hills, and valleys. The course travels under a railroad trestle and between two-story brick buildings. Tape recordings of the noise were made on the road course with all emergency devices sounding. Additional measurements were made at the engine pump panels while the engine was delivering water to a hose at 150 and 200 psi.

The results of the noise analyses showed that the engine riding positions had mean levels that ranged from 94 to 96 dB(A) in the five different positions. The ladder trucks were more variable in the noise encountered in the different riding positions, with means ranging from 77 to 105 dB(A). The quietest ride was in the tiller cab and the loudest in the jumpseat, next to the engine. Noise exposure at the pump panel was measured at 89 and 93 dB(A) for the 150- and 200-psi settings. It is these latter noise exposures that could last for longer times, depending on the amount of time the engine is required to deliver water at the fire scene.

Ambulance Survey. Noise levels inside and outside an ambulance were investigated in another NIOSH health hazard evaluation.[9] In this survey, investigators varied the location of the electronic siren speaker from the roof of the ambulance cab to the center of the front bumper. Measurements were made inside the vehicle near the driver's seat and in the patient compartment. Additional measurements were made 10 and 100 feet downrange from the stationary vehicle. Noise measurements inside of the ambulance revealed that the grille location for the siren speaker yielded the lower exposure levels in the cab and patient compartment. Reductions of 16 - 22 dB(A) were seen in the driver's cab and reductions of 8 - 15 dB(A) were found in the patient compartment when comparing the grille location to the light bar location on the roof of the ambulance. Closure of the cab's windows afforded an additional 7- to 13-dB(A) reduction in noise exposure to patient and driver. The effect of siren location on warning ability downrange from the ambulance was minimal, with only a 5-dB(A) difference in the two test conditions. The slightly higher levels were associated with the siren speaker mounted in the grille area of the ambulance.

Memphis Fire Department Survey. In 1985 NIOSH was requested to investigate the noise levels affecting the firefighters of the Memphis, Tennessee, Fire Department assigned to the Memphis Airport.[33,34] The concern was the influence that a major air freight operation headquartered in Memphis might have on the noise exposure of firefighters assigned to the airport fire stations. One of the stations is located on the airport grounds, whereas the other is adjacent to the airport property. Noise dosimetry was conducted over two consecutive 24-hour shifts at the two fire stations. To put the results into some perspective, three additional fire stations in Memphis, away from the airport, were included in the evaluation.

The results of the noise dosimetry showed that firefighters assigned to the airport stations did not have statistically different noise exposures than the firefighters assigned to the other three stations. Of the 48 noise samples collected during the survey, all of the 24-hour dose measurements were below the OSHA PEL. Close inspection of the dosimeter record during emergency responses revealed noise levels reaching 109 dB(A) for a 1-minute period. The length of the emergency run and the number of runs in a workshift thus have a great influence on the TWA noise exposure for a firefighter.

Pittsburgh and Hamilton Fire Department Survey. In 1988, NIOSH received two requests to evaluate noise levels in fire departments in Pittsburgh, Pennsylvania, and Hamilton, Ohio.[36,37] Both of these requests concerned noise exposures to firefighters during the routine operations of the fire service. Because of the large size of the Pittsburgh Bureau of Fire, only half of the 36 fire stations in Pittsburgh were investigated by NIOSH. However, all six of Hamilton's fire stations were surveyed. Both departments were surveyed with noise dosimetry, and a spectral analysis of the vehicles was done.

The Bureau of Fire in Pittsburgh has daywatch and nightwatch shifts of 10 and 14 hours, respectively. An 8-hour period of the daywatch was surveyed on two consecutive days at each of the 18 fire stations included in the evaluation. The 8-hour dose measurements were linearly interpolated to the 10-hour and 14-hour work shifts. Inspection of the 170 dosimeter records revealed that only three engine companies exceeded a 100% noise dose when the OSHA evaluation criterion was used. One company was involved with driver's training on a new tiller truck, one had pump testing and a fire run, and the third company was identified as the busiest engine company in the city during the survey. Generally, the dosimeter levels for the 10- and 14-hour shifts were less than 50% of the daily allowable noise dose at all of the companies surveyed.

The noise levels from 52 of the Pittsburgh Bureau of Fire vehicles were audiotaped while the vehicles were driven on a road course set up specifically for the survey. Median dB(A) sound levels were calculated from the one-third octave-band analyses for different riding positions (cab, jumpseat, tiller, or tailboard) and for warning signal conditions (on or off). The results of these calculations are shown in Table 1.

The Hamilton Fire Department hazard evaluation used a data collection strategy similar to that used in Pittsburgh, except that the dosimeter measurements were made over the entire 24-hour workshift. A total of 40 full-shift dosimeter samples were col-

TABLE 1. Median A-weighted Decibel Levels for Pittsburgh Bureau of Fire Vehicles at Different Riding Positions with Warning Signals On and Off

Vehicle Model	Number Tested	Driver		Passenger	
		Warning On	Warning Off	Warning On	Warning Off
1986 Pierce Engine	7	99.8	88.1	104.0[A]	100.5[A]
1988 Pierce Rear Mount Truck	9	100.0	89.6	105.2[A]	101.9[A]
1988 Pierce Tiller Truck	3	101.8	91.3	105.9[A] 93.4[B]	100.7[A] 83.4[B]
1989 Pierce Lance Engine	1	96.2	89.2	89.8[A]	85.7[A]
1976 Seagrave Tiller Truck	1	105.9	89.4	119.4[A]	101.0[A]
1977 Seagrave Rear Mount Truck	1	108.2	91.1	109.2[A]	99.2[A]
1982 Seagrave Front Mount Truck	1	104.0	—	101.9[A]	—
1978 Brockway Engine	4	106.1	87.9	103.4[C]	82.7[C]
1981–82 Grumman Engine	10	106.0	83.7	103.2[C]	79.8[C]
1973–74 Mack Engine	4	113.2	84.4	98.9[C]	85.0[C]
1984 Thibault Rear Mount Truck	4	99.4	85.9	102.1[A]	93.2[A]
1986 Dodge Wagons of Chevrolet Suburban	7	101.4	—	97.8[D]	—

A = jumpseat riding position; B = tiller riding position; C = tailboard riding position; D = backseat.

lected over two consecutive days at the six fire stations. Only one engine company had a median noise dose that exceeded 50%, which included one firefighter who registered a 118% dose on one day. This engine company used a 1975 Pirsch engine, which was one of the two loudest vehicles in the department. The other engine, truck, and EMS companies had 24-hour dose exposures of less than 50%. The spectral analysis of the vehicles showed that the diesel-powered vehicles were generally louder than gasoline-powered ones, with the maximum sound energy, in excess of 120 dB, occurring at the 125-Hz third-octave band. There were also noise measurements that confirmed that newer vehicles in the department, which had been purchased stipulating noise-reduction specifications, were in fact quieter than their older counterparts. A newer EMS squad had sound levels 20 dB(A) lower than those of an older vehicle.

Conclusions Regarding Noise Exposures. A few fire departments have begun hearing conservation programs for their firefighters, which include the measurement of vehicle noise during code 3 responses. Two of the departments, the Anaheim Fire Department in California and the Phoenix Fire Department in Arizona have reported the results of the efforts.[8] The Phoenix Fire Department has recorded noise levels up to 110 dB during code 3 operations. However, as in the NIOSH surveys, the dosimeter data were generally of low intensity, varying from 57–85 dB(A) as a TWA. The Anaheim Fire Department's noise measurements have been 97 to 103 dB for open-cab trucks during code 3 operations. The Anaheim department has purchased rear-engine, enclosed-cab vehicles and found that the noise levels are reduced to 85 dB during emergency responses.

It can be seen that most of the reported noise data for the fire services can be characterized as having the potential for intense noise exposures during emergency operations, with some levels exceeding 120 dB for short times. However, full-shift noise dosimetry consistently documents lower noise levels, generally less than any relevant evaluation criteria in use. Inspection of the hearing loss data has revealed consistent findings despite the contradictory noise data.

OCCUPATIONAL HEARING LOSS RESEARCH

Indications that occupational noise may be related to the hearing ability of firefighters and EMS personnel began to be reported in the 1970s. A graduate degree thesis describes a temporary theshold shift study conducted on 10 male fire rescue personnel following exposure to siren noise.[4] The researcher reports that the subjects showed a statistically significant, but not clinically significant, difference in their hearing ability following siren exposure. Two studies conducted in California looked at firefighters' audiograms.[22,23] In a group of 89 Department of Forestry personnel, the hearing tests showed that, on average, firefighters have poor hearing regardless of the age of the firefighter or the years of fire service. However, the unavailability of pre-employment audiograms made it difficult to determine when the permanent threshold shift occurred. The second investigation found that 750 firefighters with the Los Angeles City Fire Department had hearing losses at the 3,000-, 4,000-, and 6,000-Hz audiometric test frequencies. Compared to general population normative data, after correction for age, hearing loss among the firefighters was more severe. The results suggested that the hearing loss measured in these firefighters was due to occupational exposures to noise.

The majority of the research on hearing loss in fire service and EMS employees has been published since 1980. The hearing levels of EMS personnel were examined in two studies reported in the early 1980s. Johnson et al. examined the hearing of 56 paramedics over a 14-year period.[12] The mean noise levels in the ambulance cabs ranged from 96 to 102.5 dB(A), depending on the up or down position of the driver's and passenger's windows. The audiometric data indicated that the ambulance paramedics appeared to lose hearing over the 14 years at a faster rate than a control population selected from non-EMS personnel at the hospital. This research was initiated because of changes that were made in siren location and increases in siren intensity which were felt to be necessary to alleviate problems with the ambulances moving through traffic.

Houston Fire Department Survey. A study of the EMS personnel of the City of Houston Fire Department reported findings similar to those of the previous study.[20] In a study population of 192 firefighters, a significant correlation between duration of siren use and hearing loss was detected. In addition, the rate of hearing loss for this group of EMS personnel was 150% of that expected in an age-matched, non-noise-exposed group in the higher audiometric test frequencies (3,000–6,000 Hz). In a second study of the Houston Fire Department, the researchers compared white and black EMS personnel.[11] Both groups of firefighters had significant hearing loss at 6,000 Hz, but the average hearing loss was consistently greater in the white participants. For the better-hearing ear, the average difference between the two groups was 3 to 5 dB, but a difference of 4 to 10 dB was measured in the poorer-hearing ear. The authors of all these studies concluded that preventive measures need to be taken to reduce the risk of increased hearing loss in EMS workers from occupational noise sources.

Newburgh Fire Department Survey. NIOSH began studying firefighter hearing loss with the health hazard evaluation of the Newburgh, New York, Fire Department.[30] This fire department had the local health department administer hearing tests to the firefighters because of concern by a number of personnel that their hearing was deteriorating. The results of the hearing tests tended to confirm this belief. NIOSH investigators were requested to repeat the audiometric screening under more controlled conditions, as well as to document noise exposures. The hearing test results showed a mean hearing loss of 62 dB HL after 30 years of fire service for a group of five firefighters over 50 years of age. Compared to audiometric data reported in a 1960–1962 National Health Survey, for the noise-sensitive audiometric test frequencies of 3,000, 4,000, and 6,000 Hz, the youngest group of firefighters had better-than-normal hearing, but the oldest group of employees had greater hearing loss than the population norms.[17] It thus appears, for this small department, that the firefighters started their career with better-than-normal hearing but lost it at a rate greater than the general population.

IAFF Surveys. NIOSH continued its evaluation of firefighter hearing loss at two annual conventions of the International Association of Fire Fighters in 1984 and 1987.[32,35] Participants at the conventions were given the opportunity to receive an audometric screening examination. A total of 419 firefighters were tested at the Cincinnati convention in 1984, and 333 were tested in Anaheim in 1987. More than 50% of the tested individuals at each survey site showed some degree of hearing loss at one or more of the test frequencies when compared to a single-frequency hearing-impairment criterion.[6] A statistically significant relationship between high-frequency hearing loss and time in the job was found for the participants in the Cincinnati convention. Although a similar relationship was not discovered for firefighters at the other convention, the hearing abilities of these people still showed a pattern consistent with excessive noise exposures. One pitfall of these kinds of surveys is a self-selction bias because the firefighters are not selected randomly or by some stratification paradigm to participate in the audiometric screening. This problem is reduced in the three additional evaluations that NIOSH has conducted in U.S. fire departments.

Memphis Fire Department Survey. The NIOSH survey of the Memphis, Tennessee, Fire Department emphasized the risk to hearing from noise emanating from nearby airport operations.[33,34] The hearing of the firefighters assigned to the two stations on or adjacent to the airport was compared to that of personnel at three other stations in the Memphis area. The nearly 200 firefighters included in the evaluation represent approximately 25% of the department's fire-suppression force. The hearing tests of the group showed the characteristic dip at the 4,000- and 6,000-HZ audiometric test frequencies, indicative of noise-induced threshold shifts. A statistical comparison of the airport firefighters to the control personnel failed to show a difference between the groups. Thus, assignment to the airport fire stations did not appear to increase the risk of hearing loss. However, after the hearing data had been corrected for aging (presbycusis) effects, a significant effect of time in the fire service was found on both groups of firefighters' high-frequency hearing.[18] For this department, the occupation of firefighting seems to pose a hazard to one's ability to hear high-frequency sounds.

Pittsburgh Fire Department Survey. Firefighters with the Pittsburgh Bureau of Fire were given audiometric screening test in conjunction with the health hazard evaluation done at this department.[36] Pittsburgh employs nearly 1000 firefighters; 424 of them were included in the audiometric test protocol because they were assigned to the fire stations where noise dosimetry was performed. The hearing test results showed a decline in the percentage of firefighters exhibiting normal hearing from 40% for personnel with less than 6 years of fire service to 0% for firefighters with more than 20 years with the department.[6] The mean hearing levels for the firefighters also showed the characteristic loss of hearing in the higher audiometric test frequencies, indicative of excessive noise exposures. The statistical analysis of the data found significant relationships between hearing loss and years of fire service, both for the noise-sensitive high frequencies as well as the lower frequencies responsible for speech recognition.

Hamilton Fire Department Survey. Finally, 90 firefighters from the 100-person department in Hamilton, Ohio, were given audiometric screening tests in conjunction with the health hazard evaluation conducted by NIOSH investigators.[37] Members of this fire department had their hearing tested previously, and 55 participants in the NIOSH study had an examination for comparison. The consistent finding of high-frequency hearing loss was seen in personnel tested in the NIOSH evaluation. The smaller group of participants who had been tested 6 years earlier had a statistically significant loss of hearing for the higher test frequencies, on the order of a 5 dB average for the group. Although this change over a short time span is not clinically significant, if it were allowed to continue at the same rate for a 30-year career in the fire service, one might expect a 25-dB occupationally induced loss of hearing, which is serious.

Additional Etiologies of Hearing Loss. Because the observed noise levels in most of the studies do not explain the amount of hearing loss measured in the firefighters, some other possible explanations can be proposed. One proposal concerns the type of noise to which firefighters are exposed. The recommendations of the Committee on Hearing, Bioacoustics, and Biomechanics for maximum exposures to pure tones or narrow bands of noise is 105 dB(A) for 15 minutes, which is 10 dB less than the recommendations for broad-band noise.[14] The sirens and air horns do contribute a narrow-band component to the overall vehicle spectra measured in the surveys, normally at 2,000 and 500 Hz, respectively. However, there is considerable noise energy in all of the octave bands. Whether or not the narrow siren and air horn component is distinct enough to cause the observed hearing losses needs to be investigated more thoroughly.

Another possibility to explain the excessive hearing losses is an interaction between noise exposures and other agents found in the firefighters' environment. It is well documented that firefighters are exposed to several toxins at the fire scene.[15,21,29] Many of these chemical agents have been found to cause synergistic effects on hearing loss when subjects are exposed to both the noise and toxic agents simultaneously.[7,16] The changes in the ear's physiology from exposure to chemicals may alter the ear so that it is unable to tolerate as much noise as a normal ear.

Another explanation may be found in the intermittency of the noise exposures. The extreme differences between the noise levels found in a fire station and on a fire truck responding to an emergency may cause the OSHA 5-dB exchange rate to underestimate the impact of the noise on firefighters' hearing. Suter posed this argument in a paper on the relationship between hearing loss and exchange rates, stating that a noise survey that uses a dosimeter which calculates the TWA noise levels with a 3-dB exchange rate would certainly yield noise levels that are higher than those measured with a dosimeter with a 5-dB rate and, thus, may better predict the amount of observed hearing loss.[27] Unfortunately, no noise survey to date in the fire service has used dosimeters with the lower exchange rate.

HEARING CONSERVATION PROGRAMS

Hearing conservation programs should be implemented by fire departments as a way to reduce the possibility that firefighters will suffer permanent noise-induced hearing loss as a result of their employment. These programs must include ways to identify areas and equipment that are potentially hazardous to hearing and then reduce the noise produced by them. Personal hearing protection devices appropriate for the fire service should be issued until the equipment and vehicle noise is lowered to nonhazardous levels. The program must also include the periodic monitoring of firefighters' hearing levels, along with a training program to educate firefighters on the effects of noise on hearing and the handicaps associated with noise-induced hearing loss. Finally, a complete record-keeping system must be in place to track the effectiveness of the department's hearing conservation program. The specific components of industrial hearing conservation programs have been described elsewhere.[5,24,28]

Several changes have been seen recently in the types of vehicles available to fire departments. Completely enclosed cabs and jumpseats are commonplace in the fire service. These enclosures help to reduce the road and siren noise that the open-design cabs allow unabated. Even air conditioning in vehicles, which allows firefighters to ride comfortably with all windows closed, has been documented to lower sound levels by 13 to 16 dB.[9] A close inspection of vehicle advertisements in fire service magazines will show that noise reduction is a concern of manufacturers. Vehicles incorporate sound-damping materials into their design in locations where shiny metal was the norm in the past. The re-engineering of fire vehicle chassis has led to the introduction of rear-engine vehicles which help to distance firefighters from the engine when riding the vehicle (Fig. 1). Quiet rides have become a positive selling point for fire trucks.

Noise surveys in the fire service have repeatedly pointed to sirens, both mechanical and electronic, and air horns as major contributors to firefighters' noise exposures. The location of these warning devices can influence the intensity of the sound that

FIGURE 1. Engine located at the rear of the vehicle.

reaches personnel on the vehicle. If there is a straight visual path from the siren to the firefighter's ear, then the intensity of the siren is maximum (Fig. 2). The vehicle's structure should be used as a shield as much as possible. Moving air horns to openings in the center of the bumper (Fig. 3A) or electronic siren speakers to the front grille (Fig. 3B) or under the front bumper (Fig. 3C) removes a direct noise path to the ears of the riders and thus reduces the amount of noise reaching the firefighter. A study conducted by insurance company researchers reported up to a 70% reduction in the sound pressure level emitted by a mechanical siren when a metal shroud was installed around the siren housing that directs the sound in a forward direction.[10]

Hearing protection devices (HPD) can be used by firefighters during noisy operations, including emergency responses, routine tests on equipment at the station, and certain fire ground operations, such as pump panel operations. Earmuffs are the logical choice of HPDs because they can be rapidly applied to fit over the wearer's ears (Fig. 4). Ear plugs are highly dependent on a good fit in the ear canal, which takes some time to achieve. The use of communication headsets that have speakers placed inside of an HPD muff and a noise-blanking microphone are being used successfully in fire departments throughout the country. The technology of active noise cancellation has been incorporated into HPDs and is being marketed as a headset for EMS to block siren noise. These devices are more expensive than traditional ear muffs and ear plugs.

Incorporating the engineering controls and the use of HPDs into a fire service hearing conservation program can help stem the tide of permanent hearing losses that result from occupational noise exposures. However, for any program to work well, it must be consistently and fairly enforced. If the fire chief visits a station during a noisy operation, then the chief either puts on hearing protection like everyone else or is removed from the noise. The officer in charge must make abiding by the rules of the hearing conservation program just as important as the other safety rules enforced on the fire ground. When the individual in charge of the program shows a great amount of enthusiasm, it has a tendency to rub off on other firefighters, which generates a positive attitude toward hearing conservation.

Although the hearing of firefighters who have had years of fire service cannot be restored to normal, it can be kept from getting worse. Also, the hearing of new recruits into the fire service can be protected from occupational noise insults when an effective

FIGURE 2. Mechanical siren placed in a location where the emitted sound has a direct path to the driver's ears.

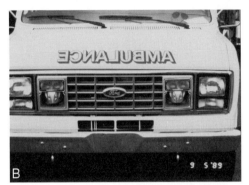

FIGURE 3. *A*, dual air horns placed in holes stamped into front bumper. Note the poor placement of the mechanical siren on the driver's side of the vehicle. *B*, electronic siren in grille-mounted speakers medial to the headlights. *C*, electronic siren speaker mounted below front bumper in center of cab.

FIGURE 4. Firefighter wearing earmuffs in preparation for vehicle movement.

hearing conservation program is implemented. Programs that work will prevent occupationally induced hearing loss.

REFERENCES

1. American Academy of Otolaryngology–Head and Neck Surgery:Otological Referral Criteria for Occupational Hearing Conservation Programs. Washington, DC, American Academy of Otolaryngology–Head and Neck Surgery, 1983.
2. American Conference of Governmental Industrial Hygienists: 1994-1995 Threshold Limit Values for Chemical Substances and Physical Agents and Biological Exposure Indices. Cincinnati, American Conference of Governmental Industrial Hygienists, 1994.
3. American National Standards Institute: Specifications for Audiometers (S3.6). New York, American National Standards Institute, 1989.
4. Beale DY: Some effects of exposure to noise on hearing threshold levels of selected ambulance drivers in the District of Columbia fire department [dissertation]. Washington, DC, Howard University Graduate School, 1974.
5. Code of Federal Regulations: OSHA. 29 CFR 1910.95. Occupational noise exposure. Washington, DC, U.S. Gov. Printing Office, Federal Register, 1992.
6. Eagles EL, Hardy WG, Catlin FI: Human Communication: The Public Health Aspects of Hearing, Language, and Speech Disorders (NINDB monograph #7). Washington, DC, Govt. Printing Office, 1968, USPHS publication 1745.
7. Fechter LD, Young JS, Carlisle L: Potentiation of noise induced threshold shifts and hair cell loss by carbon monoxide. Hearing Res 34:39–47, 1988.
8. Federal Emergency Management Association: Fire & Emergency Service Hearing Conservation Program Manual. Washington, DC, Federal Emergency Management Agency, United States Fire Administration,1992, report FA-118.
9. Flesch JP, Tubbs RL: Health hazard evaluation: Cincinnati, National Institute for Occupational Safety and Health, 1985, HHE report 84-493-1583.
10. Heany M: Effect of sound focusing device on reducing fire engine siren noise. Presented at 1994 American Industrial Hygiene Conference and Exposition, Anaheim, CA, May 23, 1994. Lynbrook, NY, Liberty Mutual Insurance Group, 1994.
11. Jerger J, Jerger S, Pepe PE, Miller RH: Race differences in susceptibility to noise-induced hearing loss. Am J Otol 7:425–429, 1986.
12. Johnson DW, Hammond RJ, Sherman RE: Hearing in an ambulance paramedic population. Ann Emerg Med 9:557–561, 1980.
13. Kryter KD: The Effects of Noise on Man. New York, Academic Press, 1970.
14. Kryter KDW, Miller JD, Eldridge DH: Hazardous exposure to intermittent and steady-state noise. J Accoust Soc Am 39:451–464, 1966.
15. Levine MS, Redford EP: Occupational exposure to cyanide in Baltimore fire fighters. J Occup Med 20: 53–56, 1978.
16. Morata TC, Dunn DE, Sieber WK: Occupational exposure to noise and ototoxic organic solvents. Arch Environ Health 49:359–365, 1994.
17. National Center for Health Statistics: Hearing Levels of Adults by Age and Sex: United States, 1960–1962. Hyattsville, MD, National Center for Health Statistics, 1965, DHEW publication PHS 79-1063.
18. National Institute for Occupational Safety and Health: Criteria for a Recommended Standard: Occupational Exposure to Noise. Cincinnati, National Institute for Occupational Safety and Health, 1972, DHEW (NIOSH) publication 73-11001.
19. Niemeier RW: Memorandum of April 13, 1995, from Division of Standards Development and Technology Transfer, to Division Directors, National Institute for Occupational Safety and Health, Centers for Disease Control and Prevention.
20. Pepe PE, Jerger J. Miller RH, Jerger S: Accelerated hearing loss in urban emergency medical services firefighters. Ann Emerg Med 14:438–442, 1985.
21. Redford EP, Levine MS: Occupational exposure to carbon monoxide in Baltimore fire fighters. J Occup Med 18:628–632, 1976.
22. Reischl U, Bair HS, Reischl P: Fire fighter noise exposure. Am Ind Hyg Assoc J 40:482–489. 1979.
23. Reischl U, Hanks TG, Reischl P: Occupational related fire fighter hearing loss. Am Ind Hyg Assoc J 42:656–662, 1981.
24. Royster LH, Royster JD, Berger EH: Guidelines for developing an effective hearing conservation program. Sound Vib 16:22–25, 1982.
25. Society of Automotive Engineers: A Study of Noise Induced Hearing Damage Risk for Operators of Farm Construction Equipment. Warrendale, PA, Society of Automotive Engineers, 1968, SAE research project R-4.

26. Suter AH: The Ability of Mildly Hearing-Impaired Individuals to Discriminate Speech in Noise. Washington, DC, U.S. Environmental Protection Agency, Joint EPA/USAF study, 1978. EPA 550/9-78-100, AMRL-TR-78-4.
27. Suter AH: The Relationship of the Exchange Rate to Noise-Induced Hearing Loss. Cincinnati, National Institute for Occupational Safety and Health, 1992, NIOSH report 9139562.
28. Suter AH, Franks JR: A Practical Guide to Effective Hearing Conservation Programs in the Workplace. Cincinnati, National Institute for Occupational Safety and Health, 1990, DHHS(NIOSH) publication 90-120.
29. Treitman RD, Burgess WA, Gold A: Air contaminants encountered by fire fighters. Am Ind Hyg Assoc J 41:796–802, 1980.
30. Tubbs RL, Flesch JP: Health Hazard Evaluation: Newburgh Fire Department. Cincinnati, National Institute for Occupational Safety and Health, 1982, HHE report 81-059-1045.
31. Tubbs RL: Health Hazard Evaluation: The City of New York Fire Department. Cincinnati, National Institute for Occupational Safety and Health, 1985, HHE report 81-459-1603.
32. Tubbs RL, Melius J, Anderson KE: Health Hazard Evaluation: International Association of Fire Fighters. Cincinnati, National Institute for Occupational Safety and Health, 1988, HHE report 84-454-1890.
33. Tubbs RL: Health Hazard Evaluation: Memphis Fire Department. Cincinnati, National Institute for Occupational Safety and Health, 1990, HHE report 86-138-2017.
34. Tubbs RL: Occupational noise exposure and hearing loss in fire fighters assigned to airport fire statioins. Am Ind Hyg Assoc J 52:372–378, 1991.
35. Tubbs RL: Health Hazard Evaluation: International Association of Fire Fighters. Cincinnati, National Institute for Occupational Safety and Health, 1991, HHE report 87-352-2097.
36. Tubbs RL: Health Hazard Evaluation: Pittsburgh Bureau of Fire. Cincinnati, National Institute for Occupational Safety and Health, 1994, HHE report 88-0290-2460.
37. Tubbs RL: Health Hazard Evaluation: Hamilton Fire Department. Cincinnati, National Institute for Occupational Safety and Health, HHE report 89-0026-2495.
38. Ward WD: Anatomy and physiology of the ear: Normal and damaged hearing. In Berger EH, Ward WD, Morrill JC, Royster LH (eds): Noise & Hearing Conservation Manual. 4th ed. Akron, OH, American Industrial Hygiene Association, 1986, pp 177–195.

ROBERT HARRISON, MD, MPH
BARBARA L. MATERNA, PhD, CIH
NATHANIEL ROTHMAN, MD, MPH

RESPIRATORY HEALTH HAZARDS AND LUNG FUNCTION IN WILDLAND FIREFIGHTERS

From the Occupational Health
 Branch
California Department of Health
 Services
Berkeley, California (RH, BLM)
 and
Department of Environmental
 Health Sciences
The Johns Hopkins University
School of Hygiene and Public
 Health
Baltimore, Maryland (NR)

Reprint requests to:
Robert Harrison, MD
Occupational Health Branch
California Department of Health
 Services
2151 Berkeley Way, Annex 11
Berkeley, CA 94704

Thousands of firefighters are exposed annually to extremely hazardous working conditions during fire suppression activities at wildland (grass, brush, and forest) fires. Each year in the United States, an estimated 80,000 firefighters are involved in 70,000 fires burning a total of 2 million acres.[41]

Although wildland firefighters share with structural firefighters an exposure to smoke, their work practices differ significantly; therefore, their exposure levels to airborne contaminants can be expected to differ. Wildland firefighters are not likely to experience the extreme acute exposures that structural firefighters may encounter while working in confined spaces. However, they may spend several days or weeks working in smoke for shifts of 12 hours or longer. Use of a self-contained breathing apparatus is not feasible, and respiratory protection consists primarily of bandannas tied over the nose and mouth.[37] Additional off-shift exposure may be incurred when fire base camps are located in areas that fill with smoke, especially during evening temperature inversions.

Wildland firefighters may be exposed to a multitude of contaminants that are the products of combustion of natural materials, including carbon monoxide, carbon dioxide, nitrogen oxides, sulfur dioxide, and other toxic gases; particulate matter of variable composition; and aldehydes, polyaromatic hydrocarbons (PAHs), and other organic compounds. Exposure to substances such as lead or herbicides that have been deposited on foliage and may become airborne in a fire also may present potential hazards.[19,40] In addition, work activities may cause suspension of ground dust and exposure to naturally occurring silica or asbestos. Chemicals such as fire-retardant materials are used to fight fires, and gasoline (which contains benzene) and other fuels are used for intentional burning. These potential exposures are associated with a range of possible acute adverse health effects, such as neurotoxic injuries, eye irritation, and upper and lower respiratory airway irritation; and chronic adverse health effects, including cancer, cardiovascular disease, airway hyperreactivity, and fibrotic lung diseases.

In contrast to the information regarding acute and chronic respiratory health effects associated with the hazardous exposures during structural firefighting, relatively limited data are available that describe the morbidity or mortality experience of wildland firefighters related to possible airborne contaminant exposures. Symptom surveys have indicated a high frequency of cough and upper respiratory irritation among wildland firefighters,[37] and substantial numbers of medical visits by firefighters for respiratory problems were observed during the 1988 Yellowstone National Park fires.[41] In the last five years, several studies have demonstrated acute changes in lung function over the course of a fire season, and there have been increasing efforts to develop practical, safe, and reliable strategies to reduce wildland firefighter smoke exposure.

AIRBORNE CONTAMINANT EXPOSURES

Until recently, few industrial hygiene studies had been carried out under wildfire or more controlled fire conditions (e.g., test burns). The studies described below provide personal monitoring data on a number of contaminants, including potential respiratory irritants (aldehydes such as formaldehyde and acrolein, respirable particulate) and other chemicals. Comparison of air concentration results across studies is made more difficult by differences in sampling times and sampling strategies. Short-term sampling is typically done when firefighters are known to be in heavy smoke environments, in an effort to assess "worst case" exposure situations. Full-shift or longer-term time-weighted averages may represent widely varying smoke conditions, including lengthy periods out of smoke, and/or a variety of work tasks, all of which are often difficult to document in the field. Shift lengths are often much longer than 8 hours but usually include travel time to the site of the fire. Current occupational exposure limits (legal and recommended) for the contaminants with sampling data are provided in Table 1.

Carbon Monoxide

Carbon monoxide exposure has been studied by using biologic monitoring, typically CO concentration in end-exhaled breath, to estimate carboxyhemoglobin (COHb) and CO air levels. In one study, 17 individuals (18.1%) were found to have a calculated COHb level above 5%, with values for three firefighters at approximately 9%.[37] The American Conference of Governmental Industrial Hygienists (ACGIH) recommends an end-of-shift COHb limit of 3.5%.[1] This study was confounded by a high prevalence of cigarette smoking; only one nonsmoker, who had spent several hours operating an engine in an area with an ambient CO level of 200 parts per million (ppm), had a COHb level above 5%. A second study found that, among 1,661 nonsmoking firefighters at 22 separate fires, 159 (10%) had COHb levels above 5%.[11] A third study during 1988 fires in Yellowstone National Park did not detect a statistically significant increase in COHb across shifts for the entire group but did find a slight increase in one crew; the highest level measured was a 4.7% COHb preshift in a smoker.[25] A fourth study, based on experimental fires of short duration (37–187 minutes), did not find evidence that firefighters were likely to experience hazardous levels of COHb.[5]

Instantaneous area CO measurements were taken with colorimetric detector tubes at a California wildfire base camp and also at the fireline during mop-up and during

TABLE 1. Occupational Exposure Limits for Components of Wildfire Smoke

Contaminant	OSHA Permissible Exposure Limit (PEL)	ACGIH Threshold Limit Value (TLV)	NIOSH Recommended Exposure Limit (REL)
Carbon monoxide	50 ppm TWA	25 ppm TWA	35 ppm TWA 200 ppm Ceiling
Sulfur dioxide	5.0 ppm TWA	2.0 ppm TWA 5.0 ppm STEL	2.0 ppm TWA
Total/inhalable particulate	15 mg/m^3	10 mg/m^3	—
Respirable particulate	5.0 mg/m^3	3.0 mg/m^3 (proposed 1994–95)	—
Crystalline silica/quartz	0.1 mg/m^3 (respirable)	0.1 mg/m^3	Lowest level feasible
Formaldehyde	0.75 ppm TWA 2.0 ppm STEL	0.3 ppm Ceiling	0.016 ppm TWA 0.1 ppm STEL
Acrolein	0.1 ppm TWA	0.1 ppm TWA 0.3 ppm STEL	0.1 ppm TWA 0.3 ppm STEL
Benzene	1.0 ppm TWA 5.0 ppm STEL	0.3 ppm TWA-skin (proposed 1994–95)	0.1 ppm TWA 1.0 ppm STEL

OSHA = Occupational Safety and Health Administration
ACGIH = American Conference of Governmental Industrial Hygienists
NIOSH = National Institute for Occupational Safety and Health
TWA = 8-hour time weighted average
STEL = Short-term (15 min.) exposure limit
Ceiling = limit not to be exceeded

the operation of gasoline-powered pumping engines, and a direct-reading CO monitoring device was used to measure CO readings at the fireline over a 45-minute period. Carbon monoxide levels ranged from 3–80 ppm, but firefighters who tended gasoline-powered pumping engines were exposed up to 300 ppm.[17] Colorimetric diffusion tubes were used for personal and area full-shift CO sampling at the 1988 Yellowstone fires; TWA levels ranged from 1.9–7.8 ppm on personal samples, up to 23.3 ppm for area measurements.[25] At a separate series of 1989 California wildfires and prescribed (planned) burns, personal time-weighted average (TWA) CO exposures were measured for 46 firefighters using colorimetric diffusion tubes during mop-up and prescribed burning (sampling times 151–696 minutes, mean 330 minutes); a mean of 14.4 ppm CO was obtained, with five employee exposures (11%) exceeding the 25-ppm, 8-hour threshold limit value (TLV).[17] An operator of a gasoline-powered pumping engine had one of the higher TWAs, at 27 ppm over a 334-minute sampling duration.

Three methods were used to assess personal exposures to CO among 25 firefighters at a single California wildfire in 1990: passive colorimetric diffusion tubes, passive electrochemical datalogging monitors, and end-exhaled air monitoring.[18] Full-shift TWA diffusion tube samples (n=51) ranged from 2–16 ppm, with a mean of 8.2 ppm; full-shift TWAs measured with dataloggers on 12 of the same firefighters ranged from 1–16 ppm with a mean of 7.8 ppm. For two individuals, a single 1-minute CO level exceeded the NIOSH recommended ceiling limit of 200 ppm; these firefighters were engaged in mop-up activities at the time. The highest 15-minute TWA was 150 ppm for one firefighter (during mop-up), but for all other individuals it was 50 ppm or less. End-exhaled air measurements showed a small but statistically significant increase in CO and percent COHb across the shifts (mean of 2.2 ppm CO or 0.39% COHb), with post-shift CO levels in end-exhaled air significantly higher than preshift values.

At the 1990 fires in Yosemite National Park, full-shift TWA CO exposures measured for a crew of 10 firefighters using colorimetric diffusion tubes had a mean of 18.3 ppm; the mean for a different nine-member crew in lighter smoke was 3.9 ppm.[26] The investigators used actual conditions of altitude, work rate, and hours exposed and the Coburn, Foster, and Kane equation for uptake of CO to estimate that average exposures should be kept below 21 ppm to keep COHb less than 5%; 30% of the first crew exceeded this level, but none of the second crew did.

Full-shift CO concentrations were measured for 20 firefighters on three successive days at a Montana wildfire using colorimetric diffusion tubes.[24] The means for each day were 1.6, 6.9, and 6.2 ppm, with individual measurements ranging from nondetectable to the two highest samples, 16 and 17 ppm.

Carbon monoxide was collected through a flame-resistant, portable sampling train into gas sampling bags for infrared spectroscopy analysis at 14 prescribed burns and two wildfires over a 2-year period.[29] Firefighters were sampled only when in visibly heavy smoke, with most sampling times in the range of 30–60 minutes. A total of 166 samples were obtained, with the majority of TWA results less than 35 ppm; one 25-minute sample exceeded the ceiling limit of 200 ppm. Colorimetric diffusion tubes were also used and found to correlate well with the gas bag samples.

In a follow-up study, 200 personal samples for CO (and other chemicals described below) were collected at 38 prescribed burns over a 3-year period using gas sampling bags and passive electrochemical dataloggers.[31] Full-shift TWA exposures were calculated by using a combination of sampled concentrations while workers were in smoke (using 2-hour samples), assumed zero exposures when in non-smoky areas, and estimated levels for some unsampled exposure time from interpollutant correlation relationships or data from contiguous sampling periods. Shifts averaged 11.5 hours, with an average of 7 hours per shift on-site during a burn. TWA exposures to CO had a geometric mean of 4.1 ppm over the full shift and 6.9 ppm while the firefighters were on site during the prescribed burn. The highest full-shift TWAs were between 35–40 ppm. Onsite CO levels were highest during direct attack (geometric mean 33.2 ppm). Eighteen peak samples during "intense" smoke exposure situations showed a geometric mean CO level of 54.3 ppm, with a maximum of 179.4 ppm; these samples were approximately 15 minutes in duration. Carbon monoxide TWA measurements were highly correlated with both formaldehyde and respirable particulate concentrations, suggesting that it may be possible in some situations to consider CO sampling as a surrogate for these contaminants. Although the continuous dataloggers had some problems with accuracy and maintenance, they provided detailed information regarding variable CO exposure during the workshift.

Engine crew members from a single fire station in Northern California were monitored for CO (and other contaminants as described below) during all fires that occurred during daylight hours over 11 days in a 2-year period; the objective was to monitor exposures during the first-response phase of fires, known as the initial attack.[30] Exposure to CO in 56 samples averaged 1.9 ppm during the 15-hour shift and 7.0 ppm at fires; the average portion of the shift spent at a fire was 2.3 hours. Firefighters reported that smoke exposure at these fires was relatively low. Samples collected during 15- to 20-minute peak exposures showed maximum CO concentrations during attack (geometric mean 12 ppm, maximum 42.2 ppm). Interpollutant correlations were limited due to small sample numbers and were not as strong as in the previous study.

Sulfur Dioxide

Air has been monitored for sulfur dioxide (SO_2) at a number of wildfires in response to concern over its irritant properties and the possibility of occurrence at wildfires in certain geographic areas. The only sampling method used to date is colorimetric diffusion tubes; these data should be considered as suggestive findings until corroborated with a method less subject to interferences from other substances. No data are available on short-term levels of SO_2.

Full-shift SO_2 concentrations at the 1988 Yellowstone fires ranged from nondetectable to 1.2 ppm, with a mean of 1.0 ppm; area measurements at three fires were 1.0, 1.8, and 1.9 ppm.[25] At the 1990 Yosemite fires, full-shift SO_2 levels for 11 firefighters had a mean of 1.4 ppm, with two levels (18%) exceeding 2.0 ppm.[26]

Full-shift SO_2 concentrations were measured for 20 firefighters on three successive days at a Montana wildfire using colorimetric diffusion tubes.[24] The means for each day were 1.0, 1.7, and 1.5 ppm, with individual measurements ranging from 0.6–3.0 ppm. Three measurements exceeded 2.0 ppm despite firefighters' reports that the smoke exposure was low to moderate.

Particulate and Silica

Total particulate exposure at wildfires and prescribed burns was determined using tared 5-μm PVC filters in 37-mm cassettes at flow rates of 1.0 to 2.0 L/minute (sampling times 206–491 minutes, mean 345 minutes), and respirable particulate was sampled using a 10-mm nylon Dorr-Oliver cyclone in front of the tared 5-μm PVC filter cassette at a flow rate of 1.7 L/minute (sampling times 77–682 minutes, mean 274 minutes). Half of the respirable dust samples were analyzed for crystalline silica content.[17] Total particulate levels reached as high as 37 mg/m³. Of the 22 total particulate samples taken at the fireline during mop-up, three (14%) exceeded 15 mg/m³ and seven (32%) exceeded 10 mg/m³. The major contribution to particulate exposure was probably from ground dust. Respirable dust exposures, although lower on average, exceeded the 5-mg/m³ inert respirable dust PEL for one firefighter, at 5.14 mg/m³ on a 229-minute sample. Of the 21 respirable dust samples analyzed for crystalline silica, five (24%) had detectable levels ranging from 1–8%. None exceeded the OSHA PEL for samples with at least 1% crystalline silica (as quartz), which is 0.1 mg/m³. One firefighter's silica exposure approached the PEL on a 324-minute sample, at 0.091 mg/m³.

At the 1990 Yosemite fires, six full-shift respirable particulate samples ranged from 0.6–1.7 mg/m³.[26] Twenty-five full-shift respirable dust samples were collected over 3 days at a Montana fire, and half of the samples were analyzed for crystalline silica.[24] The mean concentration of 24 of these samples was 0.37 mg/m³; the last sample was 4.3 mg/m³ with no apparent explanation as to why this concentration was high. Only two samples contained measurable amounts of crystalline silica; one quartz concentration was 0.35 mg/m³ (the same sample that measured 4.3 mg/m³ for respirable dust), the other was 0.04 mg/m³.

In a study of 38 prescribed burns over a 3-year period, 338 respirable particulate samples (using 2-um Teflon filters and cyclones) were collected.[31] The geometric mean exposure to respirable particulate was 0.63 mg/m³ over a workshift that included travel time, and it was 1.0 mg/m³ onsite during burns; only about 3% of the exposure averages during burns were above the 5 mg/m³ OSHA PEL. Peak exposure samples (approximately 15 minutes) ranged from 0.88–37.11 mg/m³ (geometric mean 7.0 mg/m³). Respirable particulate exposure was greatest during attack activities and correlated well with CO, formaldehyde, and acrolein concentrations.

In the Northern California study of engine crews discussed above, 45 respirable particulate samples were collected during 11 wildfires over 2 years.[30] Geometric mean respirable particulate exposure was 0.24 mg/m³ during the workday and 1.07 mg/m³ at fires. The highest geometric mean exposure to respirable particulate occurred during mobile attack (2.49 mg/m³), with short-term peak exposure measurements ranging up to 6.88 mg/m³ (during engine work).

Polyaromatic Hydrocarbons

Twenty personal exposure measurements for polyaromatic hydrocarbons were collected during wildland fireline activities using a 37-mm Teflon filter cassette backed up by an XAD-2 sorbent tube at a flow rate of 2.0 L/minute (sampling time 206–491 minutes, mean 345 minutes).[17] Twelve individual PAH compounds were detected in the 20 samples. All were less than 1 $\mu g/m^3$, making the total PAH levels well below the OSHA PEL of 200 $\mu g/m^3$. Area samples for PAHs were collected similarly at the 1988 Yellowstone fires; of the 17 PAHs analyzed, only acenaphthene was found in all samples, and fluorene and naphthalene were found in some samples.[25] Most concentrations were less than 2 $\mu g/m^3$, except four naphthalene concentrations were in the 2 to 4 $\mu g/m^3$ range; NIOSH investigators considered these concentrations to be low to trace levels. Five personal samples at the 1990 Yosemite fires contained low concentrations of particulate-bound acenaphthene, anthracene, and naphthalene and gaseous acenaphthene, anthracene, fluoranthene, and benzo(b)fluoranthene.[26]

A biologic monitoring study was conducted to determine if polycyclic aromatic hydrocarbon-DNA adducts are elevated in the white blood cells (WBC) of wildland firefighters.[32] Elevated PAH-DNA adducts have been detected in a variety of occupations in which PAHs are present. Blood samples were collected from 47 firefighters early in the 1988 fire season in Northern California and 8 weeks later, toward the end of an active season. No differences were observed in adduct levels over this time, consistent with the relatively low levels of PAHs detected by personal monitoring described above.

Aldehydes

Personal sampling for aldehydes was conducted at wildfires and prescribed burns using XAD-2 sorbent tubes impregnated with 10% 2-(hydroxymethyl) piperidine (sampling times 144–692 minutes, mean 289 minutes), with qualitative and quantitative analysis performed for specific aldehydes.[17] The aldehyde screen identified four aldehydes in smoke: formaldehyde, acetaldehyde, furfural, and acrolein, with a total of 30 of 35 individual samples having detectable levels of one or more aldehydes. Acrolein could not be quantified in this study due to a problem with the analytic technique. All aldehyde levels were well below OSHA limits; however, TWA formaldehyde levels for two firefighters, exposed at 0.41 and 0.42 mg/m^3 for sampling durations of 172 and 226 minutes, were above the ACGIH TLV-ceiling of 0.37 mg/m^3 (0.3 ppm).

The same method was used to monitor area aldehyde concentrations at the 1988 Yellowstone fires; formaldehyde was the only aldehyde detected in the four samples, at levels too low to reliably quantify.[7] At the 1990 Yosemite fires, the five full-shift samples that were taken showed low concentrations of acetaldehyde, formaldehyde, acrolein, and furfural.[26] Full-shift personal monitoring for aldehydes by the same method was conducted over 3 days at a Montana fire.[24] The air concentrations of aldehydes ranged from trace amounts to 0.10 mg/m^3 for acetaldehyde and formaldehyde, from below the detection limit to trace amounts for acrolein, and from below the detection limit to 0.04 mg/m^3 for furfural; these are all well below OSHA and ACGIH limits.

Exposure to formaldehyde and acrolein was assessed during 14 prescribed burns and two wildfires over 2 years.[29] Sorbent tubes were impregnated with 2,4-dinitrophenylhydrazine (DNPH) dissolved in acidified acetonitrile, and samples were collected at a flow rate of 0.2 L/minute. Direct-reading passive formaldehyde dosimeter badges were of limited utility. A total of 150 formaldehyde samples were obtained, with nine TWAs exceeding 2 ppm and one sample above 3 ppm (sample durations mostly 30–60 minutes). The 132 acrolein exposures ranged up to 0.46, with four exceeding the 0.3-ppm and 15-minute STEL.

In a follow-up study over a 3-year period, 397 samples for formaldehyde were collected at 38 prescribed burns.[31] The geometric mean exposure to formaldehyde was 0.047 ppm over a workshift that included travel and off-site time and was 0.075 ppm onsite during burns. The highest exposure averaged 0.39 ppm during the workshift. Peak exposure levels of formaldehyde (approximately 15 minutes in sampling duration) ranged from 0.013–1.456 ppm (geometric mean 0.468 ppm), with samples on 11 firefighters exceeding the recommended ceiling limit of 0.3 ppm. Mean formaldehyde exposures were greatest during the attack phase. Acrolein TWAs in 140 samples had a geometric mean of 0.009 ppm during the workshift and 0.015 ppm during burns. Acrolein was difficult to measure due to technical reasons, and many of the TWAs had portions of time where the exposures were estimated via interpollutant regressions. Acrolein concentrations were also highest during the attack phase (geometric mean 0.062 ppm) and in general were highly correlated with formaldehyde concentrations.

Sixty-two formaldehyde and acrolein exposures were measured during 11 wildfires over a 2-year period in Northern California.[30] Formaldehyde exposures had a geometric mean of 0.009 ppm during a workshift and 0.039 ppm at fires, while the geometric mean for acrolein was 0.002 during a workshift and 0.010 ppm while on the fireline. Peak exposures (approximately 15 minute samples) ranged up to 0.339 ppm for formaldehyde and up to 0.066 ppm for acrolein. Both formaldehyde and acrolein concentrations were highest during the attack phase.

Benzene

Benzene exposure of 16 firefighters using drip torches was monitored using organic vapor passive diffusion monitors;[17] 14 had levels at or below 0.07 ppm, one was at 0.8 ppm, and one exposure level was 0.5 ppm (the OSHA PEL for benzene is 1.0 ppm, and in 1994–95 ACGIH proposed a revised TLV of 0.3 ppm). Charcoal tubes were used to assess exposures to benzene during 38 prescribed burns over a 3-year period.[31] TWA exposures were calculated based on 349 individual samples ranging from approximately 15 minutes to 2 hours. The geometric mean exposure to benzene was 0.016 ppm over the workshift and 0.028 ppm during burns. Highest shift-average exposures were less than 0.07 ppm, with the highest exposure measurements during sawing (geometric mean 0.091 ppm) and attack/lighting (0.109 ppm).

Benzene exposure was measured in 62 samples during 11 wildfires in Northern California over 2 years, with a geometric mean TWA exposure of 0.003 ppm during a workday and 0.017 ppm at the fires.[30] Peak exposure samples ranged up to 0.082 ppm (during mobile attack), with a geometric mean of 0.035 ppm.

Summary of Airborne Contaminant Levels

Available exposure data suggest that when wildland firefighters spend long periods working in visible smoke, they may be at risk of exceeding regulatory and/or recommended full-shift occupational exposure limits for carbon monoxide, particulate (total or respirable) and aldehydes such as formaldehyde and acrolein. On a short-term basis, the contaminants of greatest concern are carbon monoxide and the aldehydes. The aldehydes have warning properties as exposures reach hazardous levels (causing eye or respiratory irritation), but CO does not. Firefighters working at a high level of physical activity, at altitude, and/or for shifts longer than 8 hours should adjust the allowable CO limits downward to provide additional protection. We do not know which tasks or conditions are associated with the higher exposures or how frequent they are among the exposed workforce. Sulfur dioxide exposure may be significant on a full-shift or short-term basis; however, monitoring with a more specific method is needed to confirm existing data. There are fewer available measurements on which to draw conclusions for other contaminants, such as crystalline silica and sulfur dioxide. PAHs do not appear to be present at significant levels.

Firefighters are at additional risk because they are exposed to multiple contaminants. It is been suggested that an "irritant exposure index" be calculated to assess combined exposure to respiratory irritants including formaldehyde, acrolein, and respirable particulate.[31] The index is made up of the sum of three terms, with measured concentration of the contaminant in the numerator and the occupational limit for that chemical in the denominator; an index totaling above one would be considered an overexposure. The data obtained during a series of prescribed burns suggest that, for firefighters working for an extended period in smoke, at least a small percentage would be overexposed to the combination of respiratory irritants. Other respiratory irritants not sampled in the study, such as sulfur dioxide or other aldehydes, could also contribute to the combined effect.

RESPIRATORY EFFECTS

Smoke inhalation can cause airway injury that is manifested by acute decrements in pulmonary function. Several studies have documented that structural firefighters may develop such decrements after routine firefighting in the absence of severe smoke inhalation. Investigators have shown transient decreases in FVC, FEV_1, and the maximal midexpiratory flow rate (FEF_{25-75}) in crews of municipal firefighters in association with acute exposure to smoke from a single fire incident.[14,22,34,35] Transient increases in airway responsiveness as measured by methacholine challenge also have been reported in association with acute exposure to fire smoke.[6,13] The use of respiratory protective equipment among a group of structural firefighters prevented the development of significant post-fire decrements in FEV_1 and FVC.[4] These studies indicate that exposure to smoke during routine structure firefighting frequently causes acute decrements in pulmonary function. The long-term effects of firefighting on lung function are more controversial. Peters and coworkers initially demonstrated a more rapid than expected annual decline in lung function in Boston firefighters,[28] but longer-term follow-up of this cohort[20,21] and the results of at least two other studies[7,10] failed to confirm this initial finding. On the other hand, Tepper and coworkers,[38] in a 6- to 9-year longitudinal study, demonstrated an almost twofold greater rate of FEV_1 decline in firefighters who did not wear respiratory equipment than in those who did. Other investigators[16,36,39] have postulated that repeated exposure to smoke may contribute to an excess annual decline in lung function among municipal firefighters compared with that in the normal population.

As discussed above, wildland firefighters may be exposed to airborne contaminants in smoke that differ from the products of combustion to which municipal firefighters may be exposed. Recent respiratory effects studies in wildland firefighters have focused on respiratory effects across a fire season (typically May through November), with the evaluation of forced expiratory volumes and airway responsiveness. Objectives in the studies performed to date have centered on the relationship between smoke exposure and acute pulmonary effects, with each firefighter serving as his or her own control in cross-season studies. These methods have largely been determined by the logistical problems in performing pulmonary function studies in this group of workers. Wildland firefighters are often difficult to locate for testing throughout the fire season, since they are called to fires on an emergent basis, often in remote locations that are relatively inaccessible to investigators. Testing on a uniform schedule is practically impossible due to variable crew schedules, and obtaining postseason tests before dispersal of seasonal firefighters may be difficult. Cross-shift pulmonary function testing is hampered by crew accessibility, cooperation of fatigued individuals, and difficult conditions in which to perform testing. Similarly, a pilot study using peak flow meters and log books to record smoke conditions, symptoms, and twice-daily peak airflow was unsuccessful due to poor participation.[9]

Cross-seasonal changes in pulmonary function and respiratory symptoms were studied in 52 wildland firefighters in Northern California.[33] Spirometry was performed on crews from six California Department of Forestry and Fire Protection fire stations early in the 1988 fire season and again 8 weeks later. Significant cross-season increases in symptoms of eye irritation, nose irritation, cough, phlegm production and wheezing were reported, with eye irritation and wheezing most strongly correlated with firefighting activity during the last 2 weeks of the study. Mean cross-season decreases in FEV_1 (-1.2%) and FVC (-0.3%) were observed, with cumulative hours of firefighting during the last week of the study most strongly correlated with the decline in FEV_1. A significant trend was seen in the relationship between hours firefighting in the last week of the study and decline in FEV_1.

Sixty-three full-time and seasonal wildland firefighters from five U.S. Department of Agriculture and Forest Service (USDAFS) Hotshot crews in Northern California and Montana were studied across the 1989 fire season (May through October).[35] Crew members underwent spirometry and methacholine challenge testing to assess the degree of airway responsiveness. The methacholine dose-response slope (DRS) was selected as the primary measure of airway responsiveness.[27] There were significant mean individual declines of 0.09, 0.15, and 0.44 L/s in postseason values of FVC, FEV_1 and FEF_{25-75}, respectively. No consistent significant relationships were observed between mean individual declines of FVC or FEV_1 and any of the covariates investigated (smoking status, history of asthma or allergies, full-time or seasonal employment status, history of upper/lower respiratory tract symptoms, crew membership). There was a significant increase in mean individual postseason DRS compared with the preseason, and this difference was not associated with any of the covariates above. Subjects with a history of lower respiratory symptoms and subjects with a history of asthma had a greater increase in level of DRS than those without such a history.

Wildland firefighters from six USDAFS Hotshot crews in Southern California were tested across the 1990 fire season (June to September).[23] Preseason and postseason spirometry was performed on 78 firefighters; crew superintendents or their foremen completed a questionnaire for each fire, including hours of firefighting and the intensity of smoke on the fireline. Crews were then stratified into three exposure categories based on the number of hours for each fire multiplied by the smoke intensity. Significant mean cross-season declines in lung function were observed in FEF_{25-75} (-2.3%) and FEV_1/FVC (-0.5%), but not in FVC and FEV_1. Dose-related declines in FEF_{25-75} were observed, with mean changes in FEF_{25-75} for low-, medium-, and high-exposure groups of 0.5%, -1.9% and -4.7%, respectively. Lung function changes were not associated with number of seasons of firefighting, days since the last fire, or age. There was no difference in mean cross-season changes in spirometric indices between those participants with and those without a history of asthma.

Cross-season spirometry and methacholine challenge testing were performed during the 1991 fire season on 63 firefighters from 5 USDAFS Hotshot crews in California, Oregon, and Montana.[9] Postseason symptoms of worsening upper and lower respiratory symptoms were positively associated with the number of days on the fireline and negatively associated with the number of days since last fighting a fire. A significant decrease in FVC and FEV_1 was observed across the fire season, but percent decline in spirometric indices was not significantly associated with days on the fireline. Likewise, no significant relationships were observed between declines in FVC and FEV_1 and any of the covariates investigated (gender, smoking status, history of asthma or allergies, history of upper/lower respiratory symptoms). In contrast to the 1989 study, no significant differences in crude and adjusted methacholine DRS were observed.

In the most recent study, 53 wildland firefighters from USDAFS and Bureau of Land Management crews in Washington and Oregon underwent preseason (March) and postseason (late November and early December) spirometry across the 1992 fire season.[3] In addition, cross-shift spirometry was obtained on 76 individuals during both prescribed burns and wildfires. Significant mean cross-shift declines in FVC (-0.065 L), FEV_1 (-0.150 L) and FEF_{25-75} (-0.496 L) were seen and remained significant after adjustment for covariates (recent respiratory infection, smoking status, history of asthma or allergies). A significant cross-season decline was observed for FEV_1 and FEF_{25-75} (-0.104 L and -0.275 L, respectively) but not for FVC. In multiple linear regression analyses, no relationship was observed between spirometric indices and respiratory symptoms. A group of ten firefighters were retested more than 8 months later (prior to the next fire season in 1993) and had increases in all spirometric indices compared with previous postseason testing.

In summary, pulmonary effects studies have demonstrated a consistent decline in airflow rates across the fire season, with limited data regarding recovery following the end of the fire season. One of two studies has shown an increase in airway responsiveness across the fire season; in that study, individuals with a history of asthma had a greater increase in level of DRS than those without such a history. However, no studies have demonstrated a significant relationship between spirometric indices and a history or asthma or allergies, and no study has reported the development of clinically significant asthma during the fire season. All studies to date have been notably limited by the absence of either quantitative or validated qualitative exposure estimates. Due to the logistical difficulties of obtaining representative exposure measurements, at least two studies have attempted to rely on the estimation of smoke exposure by firefighters themselves. However, the heterogeneous exposures, work activities, and relatively short-term duration of many smoke exposures may make this technique unreliable. Nonetheless, the presence of respiratory irritants in wildland fire smoke (e.g., aldehydes, particulate) may be responsible for the cross-season decreases in airflow that have been observed, particularly if postseason measures reflect relatively acute effects following short-term, higher-dose exposures during the preceding several weeks. At least one study[33] has suggested that cumulative hours of firefighting during the last week prior to postseason testing most strongly correlated with decline in FEV_1, and a limited follow-up of wildland firefighters has demonstrated apparent improvement in airflow several months following cessation of smoke exposure.[3] On the one hand, unlike municipal firefighters who may be exposed to smoke on a year-round basis, many wildland firefighters are primarily exposed to smoke during the hot, dry summer months. On the other hand, many wildland firefighters are now engaged in prescribed burning activities year-round and also respond to many structural fires in the rural-urban interface. Clearly, longitudinal studies will be necessary to estimate the cumulative effect of repeated exposure to wildland fire smoke on chronic pulmonary dysfunction. Longer-term studies would also help to define other health outcomes by compiling a database of exposure information, past medical history, smoking history, and work history.

RECOMMENDATIONS FOR AN OCCUPATIONAL HEALTH PROGRAM FOR RESPIRATORY HAZARDS

Components of a medical surveillance program for wildland firefighters are listed in Table 2.

Exposure Monitoring

Available exposure monitoring data suggest that the most important contaminant for employers to monitor during fires may be carbon monoxide. Peak levels of CO exceeding 200 ppm should be identified so that firefighters can be repositioned to safer locations; there are no other feasible options for control. The electrochemical dataloggers with alarm settings are well suited for this task; however, proper calibration and maintenance must be provided. Passive colorimetric diffusion tubes are a useful way to measure full-shift CO exposures in larger numbers of workers. Since CO concentrations have been found in some studies to correlate well with other smoke components, control of CO exposures may also provide protection from overexposures to other substances.

Research industrial hygiene studies should continue to build upon the data gathered at wildfires and prescribed burns, including measurement of concentrations of multiple contaminants. Given the difficulties associated with taking industrial hygiene measurements in the wildfire environment, it is possible that the most severe conditions have never been monitored. Further identification of the job tasks, fire conditions, and other factors associated with higher exposures would be useful. A continued effort is needed to obtain a statistically representative estimate of the mean and distribution of exposures to the major contaminants through random selection of subjects over a wide range of job classifications and fire conditions.

Exposure Reduction, Including Respiratory Protection

In most cases, the options for reducing exposure to wildland firefighters are limited to administrative controls and respiratory protection. Wherever hazardous exposures are identified through air monitoring or observation of acute health effects among firefighters, crews should be repositioned or rotated to reduce exposure to heavy smoke. Limits to extended shifts may be necessary in some instances.

The use of respiratory protection by wildland firefighters is complicated by the multicontaminant nature of the exposure, especially the presence of CO, a toxic gas without warning properties. Air-purifying respirators are available that use filters and chemical cartridges to remove particulate, organic vapors such as aldehydes, and acid gases such as SO_2. Existing product materials could be adapted if necessary to address concerns such as fire hazard or visibility. However, air-purifying respirators cannot remove CO, which would remain a potential hazard if other contaminant exposures were controlled. The self-contained breathing apparatuses used by structural firefight-

TABLE 2. Components of a medical surveillance program for wildland firefighters

---
Preplacement and periodic questionnaire
Preplacement and periodic spirometry (including FVC, FEV_1, FEF_{25-75})
Respiratory evaluation, as needed for acute or increased pulmonary symptoms (wheezing, cough, shortness of breath)
Clinical follow-up for episodes of asthma, bronchitis, or bronchospasm
Computerized surveillance database with record of demographic, medical, work history and spirometry data
Routine review of workers' compensation data for incidence of respiratory disease
---

ers are not feasible in the wildland fire environment. Existing air-purifying respirators could be used for exposure control in combination with continuous CO monitoring, for example, with electronic instruments that measure CO and activate an audible alarm when a threshold level is exceeded. One researcher has been pilot-testing a prototype full-facepiece cartridge respiratory with a CO sensor and alarm mounted in the facepiece.[12] This approach may prove useful in providing protection from particulate and organic chemical exposures in environments where CO is not present at hazardous levels.

Medical Surveillance

Fire protection agencies should establish comprehensive occupational health surveillance programs for wildland firefighters. Medical surveillance should include baseline assessment of respiratory function, with administration of a questionnaire to ascertain demographic information, medical history, smoking habits, job history, past occupational exposures, and any current or past work-related health problems. Respiratory symptoms may be assessed by means of standard questions modified from those prepared by the American Thoracic Society Epidemiology Standardization Project.[8] FEV_1, FVC, and FEF_{25-75} should be measured on a spirometer that meets American Thoracic Society performance criteria,[2] and testing should be performed by a NIOSH-trained technician. Periodic respiratory surveillance should then be conducted, on at least an annual basis, and should again include a questionnaire to ascertain intervening respiratory problems as well as repeat spirometry, preferably using the same equipment and techniques. In addition, a respiratory evaluation should be performed as needed for increased respiratory symptoms, particularly following acute smoke inhalation with symptoms of wheezing, chest tightness, or shortness of breath. Careful clinical followup should be instituted for individuals with episodes of acute bronchitis or bronchospasm following smoke exposure, as persistent reactive airway disease may occur following acute irritant chemical or particulate exposure. A surveillance database should be maintained on all wildland firefighters with a record of baseline, acute follow-up and routine periodic respiratory surveillance data (including spirometry data). Routine review of workers' compensation records should also be performed to assess the incidence of respiratory disease requiring medical treatment or lost work time.

Training

Wildland firefighters should receive training in the potential health effects associated with exposure to substances known to be present in their work environments. In addition to the components of fire smoke, hazard communication regarding contaminants of ground dust and chemicals such as fire retardants should be conducted. Available information regarding high-risk activities and conditions and possible means for reducing exposures should be discussed with employees. Training in the use of simple monitoring techniques such as carbon monoxide diffusion tubes or electrochemical dataloggers should be considered.

Research

Additional research is needed in several areas involving the respiratory health of wildland firefighters, including epidemiologic studies to evaluate the health outcomes associated with potential airborne contaminants, particularly chronic pulmonary effects; more suitable respiratory protection; whether adjustment of occupational health standards is necessary to protect from risk associated with unusual work schedules; use of adjusted carbon monoxide exposure limits to address the added risk from working at high altitudes, high temperatures, and a heavy work rate; and assessment of the potential for synergistic effects from multicontaminant exposures.

REFERENCES

1. American Conference of Government Industrial Hygienists: Threshold limit values for chemical substances and physical agents and biological exposure indices. Cincinnati, ACGIH, 1994.
2. American Thoracic Society: Standardization of spirometry—1987 update. Am Rev Respir Dis 136: 1285–1298, 1987.
3. Betchley C, Koenig JQ, van Belle G, Checkoway H: Pulmonary function and respiratory symptoms in forest firefighters [draft report]. University of Washington, Seattle, 1995.
4. Brandt-Rauf PW, Cosman B, Fleming Fallon A, et al: Health hazards of firefighters: Acute pulmonary effects after toxic exposures. Br J Ind Med 46:209–211, 1989.
5. Brotherhood JR, Budd GM, Jeffery SE, et al: Firefighters' exposure to carbon monoxide during Australian bushfires. Am Ind Hyg Assoc J 51:234–240, 1990.
6. Chia KS, Jeyaratnam J, Chan TB, Lim TK: Airway responsiveness of firefighters after smoke exposure. Br J Ind Med 47:524–527, 1990.
7. Douglas DB, Douglas RB, Oakes D, Scott G: Pulmonary function of London firemen. Br J Ind Med 42:55–58, 1985.
8. Ferris BG: Epidemiology Standardization Project: Respiratory questionnaires. Am Rev Respir Dis 118:7–53, 1978.
9. Harrison RJ, Mendonca A, Osorio AM: Lung function in wildland firefighters [report submitted to the United States Department of Agriculture]. Berkeley, CA, California Department of Health Services, 1994.
10. Horsfield K, Guyatt AR, Cooper FM, Buckman MP: Lung function in West Sussex fireman: A four year study. Br J Ind Med 45:116–121, 1988.
11. Jackson G, Tietz JG: A preliminary analysis of employee exposure to carbon monoxide on wildfires and prescribed burns. United States Department of Agriculture Forest Service Equipment Development Center, Missoula, MT, 1979, project 7551-2219.
12. Johnson J: personal communication. Lawrence Livermore National Laboratory, 1995.
13. Kinsella J, Carter R, Reid WH, et al: Increased airway responsiveness after smoke inhalation. Lancet 337:595–597, 1991.
14. Large AA, Owens GR, Hoffman LA: The short term effects of smoke exposure on the pulmonary function of firefighters. Chest 97:806–809, 1990.
15. Liu D, Tager IB, Balmes JR, Harrison RJ: The effect of smoke inhalation on lung function and airway responsiveness in wildland firefighters. Am Rev Respir Dis 146:1469–1473, 1992.
16. Loke J, Farmer DO, Matthay RA, et al: Acute and chronic effects of fire fighting on pulmonary function. Chest 77:369–373, 1980.
17. Materna BL, Jones JR, Sutton PM, et al: Occupational exposures in California wildland fire fighting. Am Ind Hyg Assoc J 53:69–76, 1992.
18. Materna BL, Koshland CP, Harrison RJ: Carbon monoxide exposure in wildland fire fighting: A comparison of monitoring methods. Appl Occup Environ Hyg 8:479–487, 1993.
19. McMahon DK, Bush PB: Emissions from burning herbicide treated forest fuels—a laboratory approach. In Proceedings of the 79th Annual Meeting of the Air Pollution Control Association. Pittsburgh, Air Pollution Control Association, 1986, p 8.
20. Musk AW, Peters JM, Wegman DH: Lung function in fire fighters, I: A three year follow-up of active subjects. Am J Public Health 67:626–629, 1977.
21. Musk AW, Peters JM, Wegman DH: Lung function in firefighters, II: A five year follow-up of retirees. Am J Public Health 67:630–633, 1977.
22. Musk AW, Smith TJ, Peters JM, McLaughlin E: Pulmonary function in firefighters: Acute changes in ventilatory capacity and their correlates. Br J Ind Med 36:29–34, 1979.
23. National Institute for Occupational Safety and Health: National Park Service, Southern California. Cincinnati, NIOSH, 1991, HETA 91-152-2140.
24. National Institute for Occupational Safety and Health: National Park Service, Gallatin National Forest, Montana. Cincinnati, NIOSH, 1992, HETA 91-312-2185.
25. National Institute for Occupational Safety and Health: National Park Service, Yellowstone National Park, Wyoming. Cincinnati, NIOSH 1992, HETA 88-320-2176.
26. National Institute for Occupational Safety and Health: National Park Service, Yosemite National Park, California. Cincinnati, NIOSH, 1994, HETA 90-0365-2415.
27. O'Connor G, Sparrow D, Taylor D, et al: Analysis of dose-response curves to methacholine. Am Rev Respir Dis 138:1412–1417, 1987.
28. Peters JM, Theriault GP, Fine LJ, Wegman DH: Chronic effect of fire fighting on pulmonary function. N Engl J Med 291:1320–1322, 1974.
29. Reinhardt TE: Firefighter exposure at prescribed burns—a study and action recommendations. Seattle, United States Department of Agriculture Forest Service Pacific Northwest Research Station, 1989.

30. Reinhardt TE, Black JSD, Ottmar R: Smoke exposure at Northern California vegetation fires: Draft final report. Seattle, United States Department of Agriculture Forest Service Pacific Northwest Research Station, 1995.
31. Reinhardt TE, Hanneman A, Ottmar R: Smoke exposure at prescribed burns: Final report. Seattle, United States Department of Agriculture Forest Service Pacific Northwest Research Station, 1994.
32. Rothman N, Correa-Villasenor A, Ford DP, et al: Occupational and dietary contribution to PAH-DNA adduct load in peripheral white blood cells of wildland firefighters. Cancer Epidemiol Biomarkers Prev 2:341–347, 1993.
33. Rothman N, Ford DP, Baser ME, et al: Pulmonary function and respiratory symptoms in wildland firefighters. J Occup Med 33:1163–1167, 1991.
34. Sherman CB, Barnhardt S, Miller MF: Fire fighting acutely increases airway responsiveness. Am Rev Respir Dis 140:185–190, 1989.
35. Sheppard D, Distefano S, Morse L, Becker C: Acute effects of routine fire fighting on lung function. Am J Ind Med 9:333–340, 1986.
36. Sparrow D, Bosse R, Rosner B, Weiss S: The effect of occupational exposure on pulmonary function: A longitudinal evaluation of fire fighters and nonfire fighters. Am Rev Respir Dis 125:319–322, 1982.
37. Sutton PM, Castorina J, Harrison RJ: Carbon monoxide exposure in wildland firefighters. Berkeley, CA, California Department of Health Services, 1990, field investigation FI-87-008.
38. Tepper A, Comstock GW, Levine M: A longitudinal study of pulmonary function in firefighters. Am J Ind Med 20:307–316, 1991.
39. Unger KM, Snow RM, Mestas JM, Miller WC: Smoke inhalation in firemen. Thorax 35:838–842, 1980.
40. Ward D, Hardy C: Emissions from prescribed burning of chaparral. In Proceedings of the 82nd Annual Meeting of the Air and Waste Management Association. Pittsburgh, Air and Waste Management Association, 1989, pp 1–22.
41. Ward DE, Rothman N, Strickland P: The effects of fire smoke on firefighters: A comprehensive study plan. Missoula, MT: United States Department of Agriculture Forest Service Intermountain Research Station, 1989.

DICK GERKIN, MD

FIREFIGHTERS: FITNESS FOR DUTY

From the Phoenix Fire Department
Phoenix, Arizona

Reprint requests to:
Dick Gerkin, MD
Phoenix Fire Department
3315 West Indiana School Road
Phoenix, AZ 85017

The position of fire department physician is mandated by standards developed by the National Fire Protection Association: NFPA 1500, Standard on Fire Department Occupational Safety and Health Program,[10] and NFPA 1582, Standard on Medical Requirements for Firefighters.[11] One of the functions of the fire department physician is to give candidates and firefighters clearances to work based on both medical and physical performance criteria. With the adoption of the Americans with Disabilities Act of 1990 (ADA), these evaluations have become more complex.[1] Understanding the concepts of essential job functions and reasonable accommodation is vital for both the physician and the fire department. The balance between public safety concerns and the protection of the individual against job discrimination must be determined often.

MEDICAL STANDARDS

Hiring

NFPA 1582, enacted in August 1992, is a national medical standard that many fire departments have adopted. It was written to replace part of NFPA 1001, Firefighter Professional Qualifications,[9] and to fulfill NFPA 1500, Section 8–1, Medical Requirements. NFPA 1582 was written about the same time that the ADA was developed. The creators of NFPA 1582 attempted to be sensitive to issues that were key in the ADA. The standard embodies the concept of essential job functions, which holds that there are fundamental job duties that the individual who holds the position must perform. The job description that is used in the hiring process must cite these essential functions. Nonessential job functions cannot be used to discriminate in the hiring process. The term *reasonable accommodation* describes any change in the work environment that enables an individual with a disability to enjoy equal employment opportunity. If candidate's not being hired is due to a perceived disability, the hiring authority must show that reasonable accommodation could not be achieved without undue hardship.

NFPA 1582 lists by organ system medical conditions that are assigned to either category A or category B. Firefighters or applicants with a category A condition are unfit for regular duty or hire. A category B condition might indicate unfitness for duty, but the physician determines if the condition results in the firefighter or applicant being unable to safely perform the functions of the job.

Category A conditions are mainly those that would result in sudden incapacitation of the firefighter, leading to a significant risk to the firefighter, coworkers, or the public. They include inadequate visual acuity, coronary artery disease, heart failure, pacemaker, recent seizure, hemophilia requiring factor replacement, and the absence of an eye. The selection of these category A conditions followed 5 years of meetings by the NFPA 1582 subcommittee and was based on a consensus of medical opinion. This subcommittee reviewed the NFPA 1001 standard and constantly referred to the duties performed by a firefighter. The members of the fire service on the committee, including chiefs, union members, and personnel officers, gave feedback regarding the nature and difficulty of the essential job tasks. When relevant clinical studies were available, which was infrequently, these were reviewed. The U.S Department of Transportation (DOT) regulations[12] were not considered during this process since the fire service currently is exempted from meeting these requirements in terms of driver training and certification. Although category A conditions are stated to be absolute, that is, not subject to interpretation by the physician, every selection decision should be carefully considered, weighing new medical information, legal precedents, and special departmental considerations.

One category A condition that may be subject to interpretation is coronary artery disease. Candidates being evaluated at least 6 months after a myocardial infarction or a procedure such as coronary angioplasty or coronary bypass surgery must meet certain criteria to be considered acceptable. Candidates must have good left ventricular function as determined by an echocardiogram or a nuclear gated scan, no significant arrhythmias, and no inducible myocardial ischemia on treadmill testing. The goal is not only to determine physical capacity but, more importantly, determine relative risk for an incapacitating arrhythmia.

Category B conditions are those that could result in either sudden incapacitation or the inability of the firefighter to perform the job safely. Some controversial category B conditions are asthma, impaired hearing, diabetes mellitus, anticoagulant therapy, and hypertension. Examples of category B conditions and possible dispositions are described in Table 1.

All decisions about hiring based on medical conditions must be based on thorough understanding of the ADA. The medical condition must prevent a candidate from safely performing the job of firefighter even with reasonable accommodation. Because firefighting involves public safety, more burden is placed on applicants to show that their hiring would not jeopardize the safety of the public. Many of the controversial medical conditions may be the subject of legal action initiated by applicants who have been refused hire. Within local legal jurisdictions, rulings have been handed down regarding these conditions. These findings may or may not be applicable the next time a dispute arises in another jurisdiction.

A fire department physician who has examined and found an applicant to have a condition that may lead to rejection of the applicant for hire should immediately discuss the situation with the appropriate hiring authority. In most cases, it is prudent to seek the medical opinion of an informed subspecialist in the appropriate area of medicine. For example, if a candidate has mild hemophilia, a hematologist who has witnessed the

TABLE 1. Examples of Category B Conditions, as Specified in NFPA 1582

Condition	Acceptable	Not Acceptable
Asthma	Mildly abnormal FEV_1 Good exercise tolerance Few medications No ER visits for bronchospasm	Moderately/severely abnormal FEV_1 Poor exercise tolerance due to wheezing Several medications ER visits for bronchospasm
Diabetes	Controlled by diet or oral hypoglycemics	Uncontrolled Insulin dependent*
Hypertension	Controlled by medication that does not affect the individual's mentation	Uncontrolled Signs of end organ damage**
Anticoagulant therapy	Drugs not causing full anti-coagulation, e.g., aspirin, persantine, low-dose heparin	Full-dose coumadin, heparin

*The risk of hypoglycemia is unpredictable. The single best indicator of risk is the history of a previous episode. However, at least one incident has occurred in which an applicant, previously refused hire because of insulin-dependent diabetes, was hired in Florida after the policy of the fire marshal's bureau was overturned based on ADA considerations. The applicant had been insulin-dependent for 5 years and took insulin every 2 hours at work and every 4 hours on her days off.
**Some signs of damage such as mildly elevated renal function tests and mild retinal disease may not impair a candidate's ability to perform. Significant visual impairment or heart failure or renal failure that results in a low exercise tolerance are reasons for exclusion.

physical demands of and the nature of injuries arising from firefighting may be able to render an opinion about the medical risk of hiring the candidate. It is not known what liability is incurred by an employer for an injury to an employee who at the time of hire was recognized to have a condition that was thought to have a significant risk of morbidity associated with the risks of firefighting.

Return to Duty

Both NFPA 1500 8-1.3 and 1582 2-5 give guidelines regarding return to duty clearance after a firefighter's prolonged absence from duty. The definition of prolonged absence is not given, but 90 days has been suggested as a reasonable length of time in the proposed NFPA 1583, Recommended Practice for Firefighter Physical Performance Assessment.

In the event that a firefighter's personal physician has released him or her back to regular duty, and the department physician disagrees with this release, a procedure must be in place to deal with this dispute. If, after discussion, the two physicians cannot agree, one course of action might be to obtain a third opinion from an informed physician agreed to by all parties involved.

Periodic Exams

Both NFPA 1500 8-1.3 and 1582 2-4 give guidelines regarding periodic medical evaluations. NFPA 1500 requires that firefighters have an annual medical evaluation by the fire department physician. In NFPA 1582, the type of medical evaluation recommended is based on the age of the firefighter. For example, for age 29 and younger, a fuller medical examination is recommended every 3 years; for ages 30–39, every 2 years, and for age 40 and older, every year. The medical examination should include a physical examination, pulmonary function testing, audiometry, vision testing, and laboratory testing such as a complete blood count, a chemistry panel, and urinalysis. A resting electrocardiogram should also be done and consideration given to performing a treadmill stress test.

The necessity and timing of treadmill testing in the fire service is unclear. It is reasonable to assume that after a certain age, it may be advisable to start periodic treadmill testing. There are no data that indicate exactly what this age should be. Some departments start testing at age 40, while others start earlier and perform more frequent testing as a firefighter ages. Some use the presence of risk factors to guide the frequency of testing. The high cost of testing and the relatively high number of false positive tests have limited the use of the treadmill test by the fire service.

PHYSICAL STANDARDS

Hiring

For some time most fire departments have had a physical fitness evaluation for hiring. These "agility" tests have been quite varied. Some departments followed guidelines of NFPA 1001. Others developed their own tests based on recommendations from either internal experts or outside consultants. The validity of many of these tests is dubious.

Any current physical ability test used for hiring must reflect critical and essential job functions. If it resembles a job task simulation, it may not necessarily be more valid, but it will appear so to those observing from a litigious point of view. The test must be validated by the hiring department even if it is was obtained from another department that performed its own validation.

The proposed NFPA 1583 gives a list of sample tasks that may be used to construct a physical ability test. Although this document has not been finalized, the tasks are similar to those in the Combat Challenge test developed by Paul Davis,[4] which, in turn, are similar to those found in a study for the Canadian fire service by Gledhill and Jamnik.[6] The tasks also resemble those used by various departments in many other tests. The reason for the similarity is obvious: the tests are based on the same critical and essential firefighting tasks.

Currently, the tasks in NFPA 1583 include a victim rescue, forcible entry and ventilation, hose advance, stair climb with load, hoisting, and a carrying evolution. There are recommended weights and measures in this document, but departments are instructed to set their own local parameters based on a job task analysis. Similarly, a minimum score for passing should be set based on a local analysis of minimum competency.

A candidate meeting a designated minimum score on this type of test is presumed physically capable of performing as a firefighter. Ranking the candidates based on this type of test or on any test can be justified only if a correlation exists between performance on the test and performance on the job. If achieving a better score on the physical ability test implies that the candidate will be a better firefighter, rank ordering can be used in this portion of the hiring process.

Return to Duty

Both NFPA 1500 and 1582 give guidelines regarding return to duty clearance after a firefighter's prolonged absence from duty. These recommendations are similar to those regarding the medical clearance. The definition of prolonged absence is not given but should be developed by the fire department and its physician. NFPA 1583 is in concordance with these documents. If a firefighter fails the physical ability test, rehabilitation should be prescribed to enable the firefighter to pass a retest and safely return to duty. The minimum score should be the same for hiring, return to duty, and periodic evaluation.

Periodic Exams

Both NFPA 1500 and 1582 recommend yearly evaluations of physical ability, and NFPA 1583 is in agreement. A firefighter who fails the test should be removed from firefighting duties, having demonstrated less than the minimum physical ability to perform the job. The fact that the firefighter was "doing a good job" before the test does not ensure that the job was being performed safely. The firefighter should be placed in a rehabilitation program to expedite a return to full duty. The department should provide alternate duty positions for firefighters undergoing rehabilitation. Firefighters must retest within 90 days and, if they fail again, may retest every 30 days.

Risks

A concern that has been raised is that of the risk of musculoskeletal injury for a firefighter who is being tested for return to duty after an absence or for an annual evaluation. Injuries such as back, shoulder, or ankle sprains have occurred while using some of these tests. The key to reducing injuries is the understanding by the involved physicians of the physical demands of firefighting. Physical rehabilitation after an injury should restore a firefighter to levels near those necessary for an athlete to return to full competition. If this level of recovery is reached, and the firefighter is thought ready to return to the demands of full duty, what better time to find out that the injury is not fully rehabilitated than in the controlled setting of a job task simulation? The alternative of an injury in the uncontrolled setting of the fireground is less acceptable.

Another risk that has been seen primarily during hiring physical ability testing is that of exertional rhabdomyolysis and acute renal failure.[3] A variable length of time after the test, the candidate has weakness, flank pain, nausea, muscle pain, and inability to hold down fluids. This is associated with laboratory findings of elevated renal function tests, potassium, and creatine phosphokinase. Intravenous rehydration is necessary and, rarely, temporary hemodialysis is required.

Several risk factors have been proposed for rhabdomyolysis, including prior strenuous physical exertion, viral illness, alcohol consumption, dehydration, and ingestions of medications that either reduce renal blood flow, such as antiinflammatory drugs and diuretics, or are sympathomimetic, such as decongestants, amphetamines, and cocaine.

Deaths associated with exertional rhabdomyolysis have occurred almost exclusively in people with sickle cell trait, a condition that predisposes the carrier to life-threatening hypoxia in the setting of the increased blood viscosity and metabolic acidosis that accompanies severe rhabdomyolysis with concomitant renal failure. Some have suggested screening athletes at risk for sickle cell trait (blacks, people of Mediterranean extraction).[2] People with sick cell trait need not avoid sports, but should participate with reasonable preparation. Exertion should be undertaken with adequate hydration and, if it a high altitude, only after acclimatization.

A report in *Morbidity and Mortality Weekly Report*[5] on the experience of the New York City Fire Department in 1988–89 revealed significant morbidity associated with the hiring process. During that time, 32 candidates were hospitalized with rhabdomyolysis, renal failure, or both. Four required hemodialysis. One death occurred.

Based on an evaluation of that problem, the Phoenix Fire Department in 1990 began to monitor candidates who had taken the hiring physical agility test, looking for adverse outcomes. In Phoenix that year, 6 candidates were hospitalized for flank pain, nausea, and weakness within 2 days of testing. All had evidence of rhabdomyolysis with impaired renal function. None required treatment other than intravenous fluids with alkalinization and supportive care.

After reviewing the measures taken in New York to limit these complications of

TABLE 2. Six Steps to Help Remedy Problems Associated With the Hiring Agility Test

1. Well before the test, all candidates view a videotape of the physical ability test that emphasizes its strenuous nature.
2. All candidates drink one liter of water or a sports drink 1 hour before the test.
3. The practice of awarding bonus points for finishing under a certain cutoff time is no longer followed.
4. All candidates are monitored for 30 minutes after the test and given fluids, and vital signs are checked.
5. Paramedics are nearby to start intravenous fluids under the direction of a knowledgeable physician for anyone with persistent nausea and vomiting, dizziness, or hypotension.
6. All candidates are given a list of symptoms, including nausea and vomiting, dizziness, dark urine, oliguria, and flank or severe muscle pain, that would mandate their contacting a physician.

the hiring agility test, the Phoenix Fire Department took six steps to remedy the problem (Table 2). Since then, no hospitalizations have occurred due to complications of the test.

REFERENCES

1. Anfield RN: Americans with Disabilities Act of 1990: A primer of Title I provisions for occupational health care professionals. J Occup Med 34:503–509, 1992.
2. Brown RJ, Gillespie CA: Sickle cell trait. Physician Sportsmed 21:80–88, 1993.
3. Clarkson P: Exertional rhabdomyolysis and renal failure. Nat Strength Condit Assoc 15:33–39, 1993.
4. Davis PO: Relationship between simulated firefighting and physical performance measures. Med Sci Sports Exerc 14:65–71, 1982.
5. Exertional Rhabdomyolysis and Acute Renal Impairment-New York City and Massachusetts, 1988. MMWR 39:751–756, 1990.
6. Gledhill N, Jamnik VK: Characterization of the physical demands of firefighting. Can J Sports Sci 17:207–213, 1992.
7. Hogan JC, Bernacki EJ: Developing job-related preplacement medical examinations. J Occup Med 23:469–475, 1981.
8. Lemon P, Hermiston R: Physiological profile of professional firefighters. J Occup Med 19:337–340, 1977.
9. NFPA 1001, Firefighter Professional Qualifications. National Fire Protection Association, Quincy, MA, 1987.
10. NFPA 1500, Standard on Fire Department Occupational Safety and Health Program. National Fire Protection Association, Quincy, MA, 1992.
11. NFPA 1582, Standard on Medical Requirements for Firefighters. National Fire Protection Association, Quincy, MA, 1992.
12. Title 49, Chapter III, Part 391, Qualifications of Drivers, Federal Highway Administration, Department of Transportation.

INDEX

Entries in **boldface type** indicate complete chapters.

Abortion, spontaneous
　male contribution to, 832
　paternal toxicant exposure and, 835
Acetaldehyde exposure, 699–700, 832, 862
Acquired immunodeficiency syndrome (AIDS). *See also* Human immunodeficiency virus (HIV)
　Ryan White AIDS Resources Emergency Act of 1990, 753
Acrolein exposure, 699, 722, 822, 832, 833
　reproductive hazards of, 833
　in wildland firefighters, 862–863
Active noise cancellation, 853
Adult respiratory distress syndrome (ARDS), 714, 715, 726, 727
Agency for Toxic Substances and Disease Registry (ATSDR), 704
Airborne contaminants, 858–864
　aldehydes, 862
　benzene, 863
　carbon monoxide, 858–860
　levels of, in wildland fires, 863–864
　particulates and silica, 861, 864
　polyaromatic hydrocarbons, 862
　sulfur dioxide, 860–861
Airborne pathogens, 755–758
　influenza, 757
　tuberculosis, 747, 755–757
Air horns, 852–854
Airway disease, reactive, 731
Airway hyperresponsiveness, 731
　from smoke exposure, 793
Airway responsiveness
　firefighting and, 792–793
　in wildland firefighters, 864–866
Airways obstruction, 790–791, 793
Aldehyde exposure, 789
　in wildland firefighters, 862
Americans with Disabilities Act of 1991 (ADA), and fitness for duty, 871–872
Ammonia exposure, 722
　reproductive hazards of, 833
Amyl nitrite, 729–730
Anoxic injury, from smoke inhalation, 713
ARDS. *See* Adult respiratory distress syndrome (ARDS)
Arterial blood gases (ABGs)
　in oxygenation evaluation, 726
　in smoke inhalation, 714
Arterial oxygenation (A-a) gradient, 726
Asbestos exposure
　carcinogenicity of, 804, 805–806
Asphyxiants. *See also* Carbon monoxide exposure; Carbon monoxide poisoning
　carbon monoxide, 722–723
　chemical, 721

Asphyxiants *(cont.)*
　simple, 721
　tissue hypoxia and, 722–723
Asthma, irritant-induced, 793
ATSDR (Agency for Toxic Substances and Disease Registry), 704
Audiometric tests, 845–846
　American Academy of Otolaryngology, 845
　NIOSH, 845
　single-frequency hearing impairment, 845–846
A-weighted decibel, 844

Back injuries, workplace risk factors for, 737, 739
Benzene exposure, 700, 822, 832, 833
　carcinogenicity of, 804–805
　reproductive hazards of, 833–835
　in wildland firefighters, 863
Biologic hazards, **747–760**
Biomechanical stressors, 740
Bladder cancer, 810–811
Blood, toxic compounds in, 700
Bloodborne pathogens, 747–755. *See also* Occupational Exposure to Bloodborne Pathogens Standard
　hepatitis B (HBV), 748–749
　hepatitis C (HCV), 749–750
　hepatitis D (delta agent), 750
　human immunodeficiency virus (HIV), 750–751
　occupational exposure to, 748
　Occupational Exposure to Bloodborne Pathogens Standard, 751–755
　prevention of disease from, 751–755
Brain cancer, 808–809
Breathing zone, personal
　measurement of, 693, 695
Burn injuries, **707–719**. *See also* Burn wound care
　depth and extent of, 708–709
　early management in, 707–710
　epidemiology of, 707–708
　immediate intervention in, 707–708
　patient evaluation in, 708–709
　psychosocial issues in, 718
　rehabilitation for, 717–718
　resuscitation in, 710–712
　rule of nines for, 708
　severity classification of, 709–710
Burn wound care, 715–717. *See also* Burn injuries
　definitive dressing in, 715–716
　early excision in, 716–717
　escharotomy in, 716
　fasciotomy in, 716
　initial dressing in, 715
　scar and contracture in, 718
　surgical management in, 716–717

877

Cancers, prevalent, 807–816
 bladder, 810
 brain, 808–809
 digestive system, 812–814
 esophageal, 814
 genitourinary system, 810–812
 hematopoietic and lymphatic systems, 809
 kidney, 810–811
 large intestine, 813
 leukemia, 809
 liver, 813–814
 lung, 816
 multiple myeloma, 810
 nonHodgkin's lymphoma, 810
 pancreatic, 814
 prostate, 811–812
 skin, 815–816
 stomach, 814
 testicular, 812
Carbon dioxide exposure, 696, 832, 833
 reproductive hazards of, 833–834
Carbon monoxide exposure, 693, 696–697, 832, 833. *See also* Carbon monoxide poisoning
 cardiovascular disease and, 821–822, 825
 measurement of, 696–697
 reproductive hazards of, 833–834
 toxicity of, 696
 in wildland firefighters, 858–860, 863, 867
Carbon monoxide poisoning. *See also* Carbon monoxide exposure
 asphyxiation from, 722–723
 hyperbaric oxygen for, 728–729
 management of, 728–729
 in pregnant patients, 728
 from smoke inhalation, 714
Carbon tetrachloride, 692
Carboxyhemoglobin, 693, 793, 834
 measurement of, 726
 in smoke inhalation, 714
Carcinogen exposures, 803–807
 asbestos, 804, 805–806
 benzene, 804–805
 diesel exhaust, 804, 806–807
 formaldehyde, 806
 lack of information on, 804
 other agents, 807
 polycyclic aromatic hydrocarbon (PAH), 806
Carcinogenic hazards, **803–818**
Cardiovascular disease, **821–825**
 exposures in firefighting and, 821–822
 health studies of, 823–824
 increased risk of, 824–825
 methodologic limitations for studying, 824
 mortality studies of, 823–824
 other studies of, 824
 physical requirements for firefighting and, 822–823
 prevention of, 825
 psychological stress and, 823
Cardiovascular disease event study, 824
Category A conditions, 872
Category B conditions, 872–873

CDC Cooperative Needlestick Surveillance study, 754–755
Chemical exposure. *See also* Airborne contaminants; Combustion product exposure; specific chemicals
 concentration ranges of, 832
 to organic chemicals, 700–703
 reproductive toxicities of, 833
 from smoke inhalation, 713–714
Chlorinated hydrocarbon exposure, 789, 791
Chlorine exposure, 722, 727. *See also* Hydogen chloride exposure
Chloroform exposure, 832
Coagulation necrosis, 716
Combat Challenge test, 874
Combustion, 691. *See also* Combustion product exposure
 of natural products, 691–692
 partial, 692
 of synthetics and PVC, 691–692
Combustion product exposure, **691–704**. *See also* Combustion and specific agents
 acetaldehyde, 699–700
 acrolein, 699
 aldehydes, 789
 assessment of, 693
 benzene, 700
 carbon dioxide, 696
 carbon monoxide, 696–697
 chemicals assessed, 701–703
 chlorinated hydrocarbons, 791
 direct measures of, 692–693
 exposure duration in, 693
 formaldehyde, 699–700
 hazardous materials measurements in, 703–704
 hydrogen chloride, 697–698, 789
 hydrogen cyanide, 698–699
 indirect (surrogate) measures of, 693
 limits: PEL, REF, STEL, IDLH, STLC, 694
 measurements of, 695–703
 nitrogen oxides, 697, 789
 other volatile organics, 700–703
 pesticides, 791
 phosgene, 722, 727, 789
 polynuclear aromatics (PNA), 701–703
 polyvinylchloride (PVC), 790, 793–794
 pulmonary effects of, 789–790
 reproductive effects of, 833–835
 sources of data in, 692–694
 standards, guidelines, guideposts, 694
 sulfur dioxide, 699, 789
 sulfuric acid, 699
Communicable disease, **746–760**. *See also* Bloodborne pathogens
 airborne pathogens, 755–758
 bloodborne pathogens, 747–755
 hepatitis A (HAV), 759
 herpes simplex virus, 759
 Lyme disease, 759
 measles, mumps, and rubella, 759
 meningococcal meningitis, 759
 resources on, 760
 Streptococcus pyogenes (toxic strep), 759

INDEX **879**

Communication headsets, 853
Contaminants, airborne, 858–864
Contracture, burn wound, 718
Coronary artery disease, 872
Corticosteroids, 731
Crisis centers, 770
Critical incident response, 768, 772
  agency responsibilities in, 786–787
  employee responsibilities in, 782–784
  officer responsibilities in, 784–786
  purpose of, 781–782
  Standard Service Model for, 772, 779–787
Critical incident stress debriefing (CISD), 768
  training for, 769–760
Cyanide exposure, 698–699, 723, 832, 833
  cardiovascular disease and, 822
  reproductive hazards of, 834
Cyanide poisoning
  antidote for, 729–730
  management of, 729–730

Decibel, A-weighted, 844
Decibel unit, 844
Delta agent (hepatitis D), 750
Depression, with burn injuries, 718
Dibenzodioxin exposure, 703
Dibenzofuran exposure, 703
Dibromochloropropane (DBCP) exposure, 832
Dichlorofluromethane exposure, 832
Dicobalt EDTA (Kelocyanor), 730
Diesel exhaust exposure, 692
  carcinogenicity of, 804, 806–807
  particulates in, 695
  reproductive hazards of, 835
  toxic exposure to, 692, 695
Digestive system cancers, 812–814
Diisocyanate exposure, 703
Dimethylaminophenol (DMAP), 730
Dioxin exposure, reproductive hazards of, 833
Disease, communicable, **746–760**
Divers Alert Network (DAN), 729

Earmuffs, 853–854
Ear plugs, 853
Ecological model, occupational health, 736
EDTA, dicobalt (Kelocyanor), 730
Employee Assistance Plan (EAP), 770, 772
Ergonomics, **735–744**
  Lumbar Motion Monitor, 737, 739
Escharotomy, 716
Esophageal cancer, 814
Exhaust. *See* Diesel exhaust exposure
Exposure limits, 694

Fasciotomy, 716
Fatigue, physical fitness and, 740–741
Firefighting
  cardiovascular studies in, 823–824
  energy costs of, 822
  exposures in, 821–822
  fitness for duty, **871–876**
  heat exposure in, 822
  physical requirements for, 822–823

Firefighting *(cont.)*
  psychological stress of, 823
  work-related injury and illness in, 735
Fitness. *See* Physical fitness
Fitness for duty, **871–876**
  medical standards for, 871–874
  physical standards for, 874–876
  risks of testing for, 875–876
Fluid resuscitation, 710–712, 726
Formaldehyde exposure, 699–700, 832, 833
  carcinogenicity of, 806
  reproductive hazards of, 833
  in wildland firefighters, 862–863
Furan exposure
  carcinogenicity of, 807
  reproductive hazards of, 833
Furfural exposure, 862

Gas inhalation. *See also* Airborne contaminants and specific agents
  injury from, 789
Genitourinary system cancers, 810–812
Genotoxic compounds, 831

Halon, 692
Hazardous materials response, 703–704
  chemical exposure in, 692
HazMat response, 703–704. *See also* Combustion product exposure; Hazardous materials response
Headsets, communication, 853
Healthy worker effect
  in cardiovascular mortality studies, 823
  in mortality studies, 797–798
  in pulmonary function testing, 796
Hearing conservation programs, 852–855
  hearing protection devices (HPDs) in, 852–854
  sirens and air horns in, 852–854
  vehicles and, 852
Hearing loss, occupational, 849–852
  in ambulance paramedics, 849–850
  early studies of, 849
  exposure intermittency and, 852
  Hamilton Fire Department survey, 851
  hearing conservation programs, 852–855
  Houston Fire Department survey, 850
  IAFF surveys, 850
  Memphis Fire Department survey, 850–851
  Newburgh Fire Department survey, 850
  noise-induced mechanisms of, 844
  Pittsburgh Fire Department survey, 851
  pure tones/narrow bands and, 851
  toxic agents plus noise and, 851–852
Hearing protection devices (HPDs), 852–854
Hearing requirements, critical, 844
Heat exposure, 822
  pregnancy outcome and, 836, 837
  reproductive hazards of, 835–837
  weight of protective equipment for, 822–823
Hematopoietic system cancers, 809–810
Hepatitis A (HAV), 759
Hepatitis B (HBV), 748–749
  exposure control for, 751–752
  postexposure evaluation of, 753–755

Hepatitis B (HBV) *(cont.)*
  postexposure prophylaxis for, 753–754
  vaccination standard for, 752–753
Hepatitis B immune globulin (HBIG), 753
Hepatitis C (HCV), 749–750
Hepatitis D (delta agent), 750
Herbicide exposure, 703
  as reproductive hazard, 832
Herpes simplex virus, 759
Hiring agility test
  complications of, 875
  remedy for complications of, 875–876
Hiring standards
  medical, 871–872
  physical, 874
Human immunodeficiency virus (HIV), 750–751
  exposure control of, 751–752
  exposure to, 751
  needlestick injuries with, 750, 754–755
  postexposure evaluation of, 753–754
  postexposure prophylaxis for, 754–755
  Ryan White Comprehensive AIDS Resources Emergency Act, 753
  zidovudine treatment for, 754–755
Hydrocarbon exposure. *See* Polycyclic aromatic hydrocarbon (PAH) exposure
  chlorinated, 789
  reproductive hazards of, 835
Hydrofluoric acid exposure, reproductive hazards of, 833
Hydrogen chloride exposure, 697–698, 722, 789, 822, 832, 833
  reproductive hazards of, 834
Hydrogen cyanide exposure, 698–699, 723, 832, 833. *See also* Cyanide exposure; Cyanide poisoning
  cardiovascular disease and, 822
  reproductive hazards of, 834
Hydroxycobalamin, 730
Hyperbaric oxygen, 728–729
Hyperthermia. *See* Heat exposure
Hypoxemia, 790
Hypoxia, tissue, 722–723

Immediately dangerous to life and health (IDLH) data, 694
Incident Management System, 770
Influenza, 757
Inhalation, smoke. *See* Smoke inhalation injury
Irritant exposure index, 864
Irritant gases, site of deposition of, 789
Irritants, respiratory, 721–722, 789, 793
  cardiovascular disease and, 822
  in wildland fire smoke, 866
Ischemic heart disease, 824
Isocyanate injury, 722, 727

Kidney cancer, 810–811

Lactated Ringer's solution (LR), 711
Large intestine cancer, 813
Leukemia, 809
Liver cancer, 813–814
Lumbar Motion Monitor, 737, 739

Lung cancer, 816
  risk of, 799
Lung function, 794–797
  in wildland firefighters, **857–868**
Lyme disease, 759
Lymphatic system cancers, 809–810

Male reproductive hazards, 831–832
Measles, mumps, and rubella, 759
Medical standards. *See also* NFPA; Standards
  for hiring, 871–872
Medical surveillance program
  for wildland firefighters, 867–868
Meningococcal meningitis, 759
Methemoglobinemia
  management of, 730–731
  in smoke inhalation injury, 724, 726
Methylene blue, 730–731
Methylene chloride exposure, 832
  carcinogenicity of, 807
Mitchell's paradigm, 770
Mortality studies, 797–799
  for cardiovascular disease, 823–824, 825
Mucus secretion, 795
Multiple myeloma, 810
Musculoskeletal injury, **735–744**
  conceptual framework for, 736–737
  ecological model of, 736, 738, 743
  ergonomic interventions for, 744
  Lumbar Motion Monitor for, 737, 739
  in other occupations, 737–739
  personal factors in, 736–737, 739, 743
  reduction of, 744
  role of physical fitness in, 739–744
  situational factors, uncontrollable in, 736–737, 739
  two-level model for controlling, 743
  workplace factors in, 736–737, 739, 743

National Fire Protection Association (NFPA). *See also* NFPA
  standards, 760, 871–874
Necrosis, coagulation, 716
Needlestick, HIV, 750, 754–755
NFPA 1583, Recommended Practice for Firefighter Physical Performance Assessment, 873
NFPA 1581, Standard on Fire Department Infection Control Programs, 760
NFPA 1500, Standard on Fire Department Occupational Safety and Health Programs, 760
NFPA 1500 Standard on Fire Department Occupational Safety and Health Programs
  fitness, 871, 873–874
NFPA 1582 Standard on Medical Requirements for Firefighters and fitness for duty, 871–874
Nitrites, for cyanide poisoning, 729–730
Nitrogen dioxide exposure, 697, 822, 832
Nitrogen oxide exposure, 722, 727, 789, 822
Noise, occupational, 844–845
  assessment of, 844
  audiometric tests with, 845–856
  evaluation criteria for, 844–846
  hearing loss and, **843–855**

INDEX   **881**

Noise, occupational *(cont.)*
  OSHA and NIOSH standards for, 844–845
  permissible exposure limit (PEL) for, 844
  reproductive hazards of, 836, 837–838
  research on, 846–849
  sources of, 843
  time-weighted average (TWA) for, 845
  worker's daily noise dose of, 845
Noise cancellation, active, 853
Noise research, occupational, 846–849
  ambulance survey, 847
  conclusions regarding, 949
  on hearing loss, 849–852
  Memphis Fire Department survey, 847–848
  Newburgh Fire Department survey, 846–847
  New York City Fire Department survey, 847
  Pittsburgh and Hamilton Fire Department survey, 848–849
  study protocol in, 846
NonHodgkin's lymphoma, 810

Occupational cancer, **803–818.** *See also* Cancers, prevalent
  carcinogen exposures and, 803–807
  carcinogens and, 807–816
  confounding data in, 816–817
  latency period for, 817
  lifestyle factors and, 816–817
  limitations of epidemiologic data for, 816
  PAH-DNA adducts and, 817–818
  prevalent cancers, 807–816
  statistical significance of research on, 817
Occupational Exposure to Bloodborne Pathogens Standard, 751–755
  employee training, 752
  exposure control, 751–752
  HBV vaccination, 752–753
  postexposure evaluation, 753–755
Occupational health
  ecological model of, 736
  future of, 743
Occupational stress, **763–787**
Organic chemical exposure, 700–703
Oximetry, pulse, 726
Oxygen, hyperbaric, 728–729

PAH-DNA adducts, 862
  in occupational cancer, 817–818
Pancreatic cancer, 814
Parkland formula, 711
Particulate exposure, 789
  measurement of, 695
  in wildland firefighters, 861, 864
Pathogens
  airborne, 755–758
  bloodborne, 747–755
Perchloroethylene, 832
Periodic exams
  medical, 872–873
  physical, 875
Permissible exposure limit (PEL), 694
  for combustion products, 694

Permissible exposure limit (PEL) *(cont.)*
  for noise, 844–845
  for reproductive hazards, 832
Pesticide exposure, 703, 791
Phosgene exposure, 722, 727, 789
Physical fitness, **735–744**
  biomechanical stressors and, 740
  fatigue and, 740–741
  intervention effectiveness, 744
  intervention program components, 741–742
  intervention program conclusions, 742–743
  intervention program outcomes, 742
  intervention studies, 741
  on the job, 739–740
Physical performance assessment, 873–876
Physical requirements for firefighting, 822–823
Polyaromatic hydrocarbon (PAH) exposure
  in wildland firefighters, 862, 864
Polychlorinated biphenyl (PCB) exposure, 703
  carcinogenicity of, 807
  reproductive hazards of, 833
Polycyclic aromatic hydrocarbon (PAH) exposure
  carcinogenicity of, 806
  DNA adducts of, 862
  reproductive hazards of, 834
Polynuclear aromatic (PNA) exposure, 701–703, 822
Polyvinylchloride (PVC) exposure
  airway hyperresponsiveness to, 731
  combustion products of, 691–692
  pulmonary effects of, 790, 793–794
  reproductive hazards of, 833
Posttraumatic stress disorder (PTSD), 764–767
  application to fire and rescue work of, 766–767
  criteria/events precipitating, 766
  definition of, 764
  diagnosis of, 764–765
  stressors in, 765
  in Vietnam veterans, 765
Prostate cancer, 811–812
Psychological and organizational assistance
  organization consultation in, 780
  personal, family, occupational issues in, 779
  services to citizens and customers in, 779–780
Psychological stress
  cardiovascular disease and, 823
  reproductive hazards of, 836
Pulmonary effects of firefighting, **789–799.** *See also* Pulmonary function; Pulmonary function testing
  acute, 790–792
  chronic, 794–797
  hypoxemia, 790
  mechanisms of acute symptoms of, 792–794
  mortality studies and, 797–799
  pulmonary function deficits, chronic, 794–797
  pulmonary function tests of, 790–793
  respiratory diseases, nonmalignant, 798–799
  thermal injury, 789
Pulmonary function
  SCBA protection of, 794–795
  in wildland firefighters, 864–866

Pulmonary function testing
  in combustion product exposure, 790–793
  healthy worker effect in, 796
Pulmonary irritants, 721–722, 789, 793, 822, 866
Purified protein derivative (PPD) test, 755–758

Reactive airway disease, 731
Reasonable accommodation, 871–872
Recommended exposure limit (REL), 694
Recommended Practice for Firefighter Physical Performance Assessment, 873
Rehabilitation, testing after, 875
Reproductive hazards, **829–838**
  of acrolein, 833
  of benzene, 833–835
  of carbon monoxide and carbon dioxide, 833–834
  of combustion products, 833–835
  epidemiology of, 829
  exposure-effect pathway in, 830
  exposure timing and, 830
  of fire environment chemical, 833
  of formaldehyde, 833
  of heat exposure, 835–837
  of hydrogen chloride/cyanide, 833
  industrial hygiene and, 832–835
  male-mediated effects in, 831–832
  mechanisms of, 830–832
  of miscellaneous toxicants, 833
  of noise exposure, 836, 837–838
  nonchemical, 835–838
  of pharmacologic agents, 831
  physical activity and, 836
  of psychological stress, 836
  spermatogenesis problems and, 831
  of sulfur dioxide, 833–834
  toxicant classification in, 830–831
  toxicant studies of, 835
Respirators, air-purifying, 867–868
Respiratory disease. See Pulmonary effects of firefighting
Respiratory effects. See also Pulmonary effects of firefighting
  of wildland fires, 864–866
Respiratory hazards
  occupational health program for, 867–868
  in wildland firefighters, **857–868**
Respiratory impairment, chronic, 790–793
Respiratory irritants, 721–722, 789, 793
  cardiovascular disease and, 822
  in wildland fire smoke, 866
Respiratory protection
  effect on mortality of, 799
  in wildland firefighters, 864, 867–868
Resuscitation, burn, 710–712
  first 24 hours, 710–711
  fluid resuscitation, 710–712
  monitoring, 711–712
  second 24 hours, 711
Resuscitation, fluid, 710–712, 726
Return to duty standards
  medical, 873
  physical, 874

Rhabdomyolysis, exertional, 875
Ringer's solution, lactated (LR), 711
Rule of nines, 708
Ryan White Comprehensive AIDS Resources Emergency Act of 1990, 753

Scar, burn wound, 718
Self-contained breathing apparatus (SCBA), 693
  cardiovascular disease prevention and, 822
  protection from cardiovascular disease with, 825
  pulmonary function protection with, 794–795
  in wildland firefighters, 867–868
Short-term exposure limit (STEL), 694
Short-term lethal concentration (STLC), 694
Sickle cell trait, 875
Silica exposure, in wildland firefighters, 861, 864
Sirens, 852–854
Skin cancer, 815–816
Small airways obstruction, 790–791, 793
Smoke exposure
  chemical asphyxiants, 721
  composition of, 695, 721
  irritants in, 721–722
  measurement of, 695
  particulate concentrations in, 695
  simple asphyxiants, 721
  studies on, 693
  wildland fire, exposure limits for, 858–859
Smoke inhalation injury, 712–715, **721–731**, 790
  anoxic injury, 713
  asphyxiants in, 722–724
  blood/urine chemistry in, 726–727
  bronchodilators for, 728
  carbon monoxide poisoning management in, 728–729
  chemical injury, 713–714
  chronic respiratory impairment from, 790–793
  corticosteroids for, 731
  cyanide poisoning management in, 729–730
  diagnosis of, 714
  disposition and prognosis in, 731
  epidemiology of, 708, 721
  evaluation and management of, 724–726
  heat and upper airway injury in, 722
  irritants in, 722
  lower airway assessment in, 728
  methemoglobinemia in, 724, 730–731
  oxygenation evaluation in, 726
  pathophysiology of, 712–713, 722–724
  pulmonary irritants in, 789
  radiation studies in, 727
  reproductive hazards and, 829
  respiratory injury in, 695, 725–726
  signs and symptoms of, 713
  thermal injury from, 712–713. See also Heat exposure
  treatment of, 714–715, 728–731
  upper airway assessment in, 727–728
Sodium nitrite, 729–730
Sodium thiosulfate, 729–730
Spermatogenesis
  heat exposure and, 836
  problems with, 831

INDEX

Standard on Fire Department Infection Control Programs, 760
Standard on Fire Department Occupational Safety and Health Programs, 760
1999 Standard on Protective Clothing for Emergency Medical Operations, 752
Standards, professional. *See* NFPA and specific named standards
Standard Service Model, 772, 779–787
  critical incident response, 781–787
  psychological and organizational assistance, 779–780
Stomach cancer, 814
Streptococcus pyogenes (toxic strep), 759
Stress, occupational, **763–787**. *See also* Standard Service Model
  alternative constructions of, 767–768
  client relationships in, 770–771
  comparative tests of models of, 772–773
  contemporary views of, 763–764
  ethical dilemmas in, 770–771
  fire service issues in management of, 773–775
  improved effectiveness of management of, 775–776
  more reasoned course for, 772
  organizational preparation and response in, 768–770
  perceptions of, 767–768
  posttraumatic stress disorder (PTSD) and, 764–767
  professional issues in, 770–771
  psychological, 823, 836
  stressors in, 765
  workplace impacts on, 768
Stressors, biomechanical, 740
Styrene, as carcinogen, 807
Sulfur dioxide exposure, 699, 722, 789, 832, 833
  reproductive hazards of, 834
  in wildland firefighters, 860–861, 864
Sulfuric acid exposure, 699

Testicular cancer, 812
Testicular function, agents affecting, 831
Thermal injury. *See also* Heat exposure
  airway, 789

Thermal injury *(cont.)*
  from smoke inhalation, 712–713
Threshold limit value (TLV), 694
  for noise exposure, 845
  for wildland firefighter CO exposure, 858–859
Time-weighted average (TWA)
  for noise exposure, 845
  for wildland firefighter CO exposure, 858–860
Toluene exposure, 832
Trichloroethylene exposure, 832
Trichlorophenol exposure, 832
Tuberculosis, 747, 755–757
  epidemiology of, 755–756
  infection control plan for, 756
  prevention of, 756–757
  purified protein derivative (PPD) test for, 755–758
  respirators for prevention of, 756
  screening programs for, 756–757
  worker education about, 756–757

Vehicles, firefighting, and hearing conservation, 852
Vinyl chloride exposure. *See also* Polyvinylchloride (PVC) exposure
  reproductive hazards of, 833
Volatile organic chemical exposure, 700–703

Wildland firefighters
  airborne contaminant exposures for, 858–864
  aldehyde exposure in, 862
  benzene exposure in, 863
  carbon monoxide exposure in, 858–860
  chemical exposure in, 857–858
  combustion product exposure in, 857
  occupational exposure limits for, 858–859
  particulate and silica exposure in, 861, 864
  polyaromatic hydrocarbon exposure in, 862
  respiratory health hazards and lung function in, **857–868**
  sulfur dioxide exposure in, 860–861

Zidovudine, 754–755

Statement of Ownership, Management and Circulation
(Required by Section 3685, Title 39, United States Code)

1. Title of publication: OCCUPATIONAL MEDICINE: State of the Art Reviews
2. Publication number: 0885114X
3. Date of filing: September 25, 1995
4. Frequency of issue: Quarterly
5. Number of issues published annually: 4
6. Annual subscription price: $86.00
7. Complete mailing address of known office of publication: 210 South 13th St., Philadelphia, PA 19107
8. Complete mailing address of the headquarters of general business of the publisher: 210 South 13th St., Philadelphia, PA 19107
9. Full names and complete mailing addresses of publisher and managing editor:
   Publisher: Virginia B. Baskerville, HANLEY & BELFUS, INC., 210 S. 13th St., Philadelphia, PA 19107
   Managing editor: Virginia B. Baskerville, HANLEY & BELFUS, INC., 210 S. 13th St., Philadelphia, PA 19107
10. Owner: HANLEY & BELFUS, INC., 210 S. 13th St., Philadelphia, PA 19107
    Linda C. Belfus, 210 S. 13th St., Philadelphia, PA 19107
    Holtzbrinck Publishing Holdings Limited Partnership, 415 Madison Ave., New York, NY 10017
11. Known bondholders, mortgagees, and other security holders owning or holding 1 percent or more of total amount of bonds, mortgages or other securities: None
12. Special rates: N.A.
13. Publication name: OCCUPATIONAL MEDICINE: State of the Art Reviews
14. Issue date for circulation data below: July-September 1995
15. Extent of nature of circulation:

	Average no. of copies ea. issue during preceding 12 months	Actual no. copies of single issues published nearest to filing date
A. Total no. copies (net press run)	2697	2500
B. Paid circulation:		
1. Sales through dealers, carriers, street vendors, and counter sales	—	—
2. Mail subscription	1364	1390
C. Total paid and/or requested circulation	1364	1390
D. Free distribution by mail	4	4
E. Free distribution outside the mail	0	0
F. Total free distribution	4	4
G. Total distribution	1368	1368
H. Copies not distributed		
1. Office use, leftovers, spoiled	1329	1106
2. Return from news agents	—	—
I. Total	2697	2500
Percent paid and/or requested circulation	99.7%	99.7%

I certify that all information furnished on this form is true and complete.

Linda C. Belfus, President